Geometric Dimensioning and Tolerancing for Mechanical Design

Gene R. Cogorno

McGraw-Hill

New York   Chicago   San Francisco   Lisbon   London   Madrid
Mexico City   Milan   New Delhi   San Juan   Seoul
Singapore   Sydney   Toronto

The McGraw·Hill Companies

CIP Data is on file with the Library of Congress

Copyright © 2006 by Gene R. Cogorno. All rights reserved. Printed in the United States of America. Except as permitted under the United States Copyright Act of 1976, no part of this publication may be reproduced or distributed in any form or by any means, or stored in a data base or retrieval system, without the prior written permission of the publisher.

1 2 3 4 5 6 7 8 9 0   DOC/DOC   0 1 2 1 0 9 8 7 6

ISBN 0-07-146070-5

The sponsoring editor for this book was Kenneth P. McCombs and the production supervisor was Pamela A. Pelton. It was set in Century Schoolbook by International Typesetting and Composition. The art director for the cover was Anthony Landi.

Printed and bound by RR Donnelley.

McGraw-Hill books are available at special quantity discounts to use as premiums and sales promotions, or for use in corporate training programs. For more information, please write to the Director of Special Sales, McGraw-Hill Professional, Two Penn Plaza, New York, NY 10121-2298. Or contact your local bookstore.

 This book is printed on recycled, acid-free paper containing a minimum of 50% recycled, de-inked fiber.

Contents

Preface

This book is written primarily for the learner who is new to the subject of geometric dimensioning and tolerancing (GD&T). The primary purpose of this book is to teach the graphic language of GD&T in a way that the learner can understand and use it in practical applications. It is intended as a textbook to be used in colleges and universities and as a training manual for corporate training programs that teach engineering, design, drafting, manufacturing, and quality professionals. This book is also appropriate for a self-study course.

The material in this book is written in accordance with the latest revision of the geometric dimensioning and tolerancing standard, ASME Y14.5M-1994. GD&T is a graphic language. To facilitate understanding, there is at least one drawing for each concept discussed. Drawings in this text are for illustration purposes only. In order to avoid confusion, only the concepts being discussed are completely toleranced. All of the drawings in this book are dimensioned and toleranced with the inch system of measurement because most drawings produced in the United States are dimensioned with this system. You should be skilled at reading engineering drawings.

Organization

The discussion of each control starts with a definition, and continues with how the control is specified, interpreted, and inspected. There is a review at the end of each chapter to emphasize key concepts and to serve as a self-test. This book is logically ordered so that it can be used as a reference text.

A Note to the Learner

To optimize the learning process, preview the chapter objectives, the subtitles, the drawing captions, and the summary. Next, review the chapter once again focusing attention on the drawings and at the same time formulating questions about the material. Finally, read the chapter completely, searching for answers to the questions.

Comprehending new information from the printed page is only part of the learning process. Retaining it in long-term memory is just as important.

To optimize the learning process an to drive the information into long-term memory, review all new information at the end of the day; review it again the next day, the next week, and the next month. Review is more than just looking at the information. Review includes rereading the material, speaking it out loud, or writing it. Some learners learn best with their eyes, others with their ears, and still others learn best by doing. Everyone learns differently, and some students may learn best by doing a combination of these activities or all three. Learners can experiment to determine their own best method of learning.

A Note to the Instructor

An instructor's guide is available. The instructor's guide includes teaching strategies, midterm examinations, a final examination, and all of the answers. Also, this book is organized in such a way that the instructor can select appropriate material for a more abbreviated course. This text can also be used as supplementary material for other courses, such as mechanical engineering, tool design, drafting, machining practices, and inspection. Using this text and the instructor's guide will greatly facilitate the administration of a course in GD&T.

Gene R. Cogorno

Acknowledgments

The author wishes to express particular gratitude to his wife, Marianne, for her support of this project and for the many hours she spent reading and editing the manuscript; also, thanks go to his son Steven, who devoted considerable time and effort toward shpaing the style of this book. The author also wishes to express his thanks to Anthony Teresi and John Jensen for their engineering expertise and editorial comments. Acknowledgments also go to the McGraw-Hill Professional staff for their technical contributions and editorial comments. A special thanks goes to James Meadows, the author's first GD&T instructor, for his guidance and support throughout the years. Finally, thanks to the American Society of Mechanical Engineers for permission to reprint excerpts from ASME Y14.5 M-1994 (R2004); all rights reserved.

Introduction to Geometric Dimensioning and Tolerancing

For many in the manufacturing sector, geometric dimensioning and tolerancing (GD&T) is a new subject. During World War II, the United States manufactured and shipped spare parts overseas for the war effort. Many of these parts were made to specifications but would not assemble. The military recognized that producing parts that do not properly fit or function is a serious problem since lives depend on equipment that functions properly. After the war, a committee representing government, industry, and education spent considerable time and effort investigating this defective parts problem; this group needed to find a way to insure that parts would properly fit and function every time. The result was the development of GD&T.

Ultimately, the USASI Y14.5–1966 (United States of America Standards Institute—predecessor to the American National Standards Institute) document was produced on the basis of earlier standards and industry practices. The following are revisions to the standard:

- ANSI Y14.5–1973 (American National Standards Institute)
- ANSI Y14.5M–1982
- ASME Y14.5M–1994 (American Society of Mechanical Engineers)

The 1994 revision is the current, authoritative reference document that specifies the proper application of GD&T.

Most government contractors are now required to generate drawings that are toleranced with GD&T. Because of tighter tolerancing requirements, shorter time to production, and the need to more accurately communicate design intent, many companies other than military suppliers are recognizing the importance of tolerancing their drawings with GD&T.

Conventional tolerancing methods have been in use since the middle of the 1800s. These methods do a good job of dimensioning and tolerancing size features and are still used in that capacity today, but they do a poor job of locating and orienting size features. GD&T is used extensively for locating and orienting size features and for many other tolerancing applications. Tolerancing with GD&T has a number of advantages over conventional tolerancing methods; three dramatic advantages are illustrated in this introduction.

The purpose of this introduction is to provide an understanding of what GD&T is, why it was developed, when to use it, and what advantages it has over conventional tolerancing methods. With this understanding of GD&T, technical practitioners will be more likely to effectively learn the skill of tolerancing with GD&T. With this new skill, they will have a greater understanding of how parts assemble, do a better job of communicating design intent, and ultimately be able to make a greater contribution to their companies' bottom line.

Chapter Objectives

After completing this chapter, you will be able to

- *Define* GD&T
- *Explain* when to use GD&T
- *Identify* three advantages of GD&T over coordinate tolerancing

What Is GD&T?

GD&T is a symbolic language. It is used to specify the size, shape, form, orientation, and location of features on a part. Features toleranced with GD&T reflect the actual relationship between mating parts. Drawings with properly applied geometric tolerancing provide the best opportunity for uniform interpretation and cost-effective assembly. GD&T was created to insure the proper assembly of mating parts, to improve quality, and to reduce cost.

GD&T is a design tool. Before designers can properly apply geometric tolerancing, they must carefully consider the fit and function of each feature of every part. GD&T, in effect, serves as a checklist to remind the designers to consider all aspects of each feature. Properly applied geometric tolerancing insures that every part will assemble every time. Geometric tolerancing allows the designers to specify the maximum available tolerance and, consequently, design the most economical parts.

GD&T communicates design intent. This tolerancing scheme identifies all applicable datums, which are reference surfaces, and the features being controlled to these datums. A properly toleranced drawing is not only a picture that communicates the size and shape of the part, but it also tells a story that explains the tolerance relationships between features.

When Should GD&T Be Used?

Many designers ask under what circumstances they should use GD&T. Because GD&T was designed to position size features, the simplest answer is, locate all size features with GD&T controls. Designers should tolerance parts with GD&T when

- Drawing delineation and interpretation need to be the same
- Features are critical to function or interchangeability
- It is important to stop scrapping perfectly good parts
- It is important to reduce drawing changes
- Automated equipment is used
- Functional gaging is required
- It is important to increase productivity
- Companies want across-the-board savings

Advantages of GD&T over Coordinate Dimensioning and Tolerancing

Since the middle of the nineteenth century, industry has been using the plus or minus tolerancing system for tolerancing drawings. This system has several limitations:

- The plus or minus tolerancing system generates rectangular tolerance zones. A tolerance zone, such as the example in Fig. 1-1, is a boundary within which the axis of a feature that is in tolerance must lie. Rectangular tolerance zones do not have a uniform distance from the center to the outer edge. In Fig. 1-1, from left to right and top to bottom, the tolerance is ± .005; across the diagonals, the tolerance is ± .007. Therefore, when designers tolerance features with ± .005 tolerance, they must tolerance the mating parts to accept ± .007 tolerance, which exists across the diagonals of the tolerance zones.

- Size features can only be specified at the regardless of feature size condition. Regardless of feature size means that the location tolerance remains the same no matter what size the feature happens to be within its size tolerance. If a hole, like the one in Fig. 1-1, increases in size, it has more location tolerance, but there is no way to specify that additional tolerances with the plus or minus tolerancing system.

- Datums are usually not specified where the plus or minus tolerancing system is used. Consequently, machinists and inspectors do not know which datums apply or in what order they apply. In Fig. 1-1, measurements are taken from the lower and left sides of the part. The fact that measurements are taken from these sides indicates that they are datums. However, since these datums are not specified anywhere, they are called implied datums. Where datums are

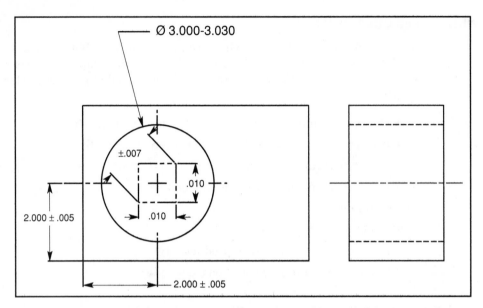

Ø 3.000-3.030

±.007

.010

.010

2.000 ± .005

2.000 ± .005

Figure 1-1 The traditional plus or minus tolerancing system. (The axis of the 3-inch hole must fall inside of the .010-inch square tolerance zone.)

implied, the designer has not indicated which datum is more important and has not specified whether or not a third datum is included. It would be logical to assume that a third datum does exist because the datum reference frame consists of three mutually perpendicular planes, but this is not specified.

When locating features with GD&T, there are three important advantages over the coordinate tolerancing system:

- The cylindrical tolerance zone
- The maximum material condition
- Datums specified in order of precedence

The cylindrical tolerance zone

The cylindrical tolerance zone is located and oriented to a specified datum reference frame. In Fig. 1-2, the tolerance zone is oriented perpendicular to datum plane A and located, with basic dimensions, to datum planes B and C. Basic dimensions have no tolerance directly associated with the dimension, thus, eliminating undesirable tolerance stack-up. The full length of the axis through the feature is easily controlled because the cylindrical tolerance zone extends through the entire length of the feature.

Unlike the rectangular tolerance zone, the cylindrical tolerance zone defines a uniform distance from true position, the center, to the tolerance zone boundary. When a .014 diameter cylindrical tolerance zone is specified about true position,

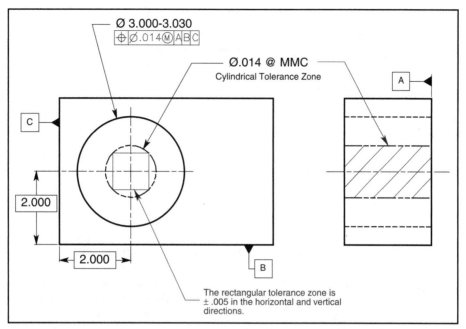

Figure 1-2 A cylindrical tolerance zone compared with a rectangular tolerance zone.

there is a tolerance of .007 from true position in all directions. A cylindrical tolerance zone circumscribed about a square tolerance zone, like the one in Fig. 1-3, has 57% more area than the square, in which the actual axis of the feature may lie.

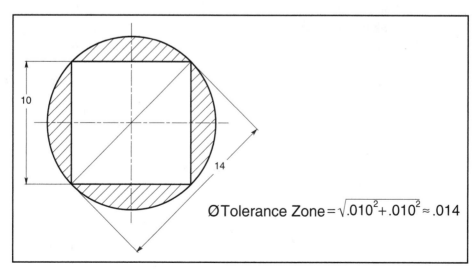

Figure 1-3 A cylindrical tolerance zone provides a uniform distance from the axis to the edge.

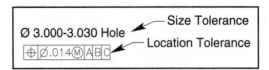

Figure 1-4 The size, size tolerance, and feature control frame for the hole in Fig. 1-2.

The maximum material condition

The maximum material condition symbol (circle M) in the feature control frame is a modifier. It specifies that as the hole in Fig. 1-2 increases in size, a bonus tolerance is added to the tolerance in the feature control frame.

The limit tolerance in Fig. 1-4 indicates that the hole size can be as small as Ø 3.000 (maximum material condition) and as large as Ø 3.030 (least material condition). The geometric tolerance specifies that the hole be positioned with a cylindrical tolerance zone of .014 in diameter when the hole is produced at its maximum material condition. The tolerance zone is oriented perpendicular to datum A and located with basic dimensions to datums B and C. As the hole size in Fig. 1-2 departs from the maximum material condition toward the least material condition, additional location tolerance, called bonus tolerance, is allowed in the exact amount of such departure. If the hole specified by the feature control frame in Fig. 1-4 is actually produced at a diameter of 3.020, the total available tolerance is a diameter of .034 of an inch.

Actual feature size	3.020
Minus the maximum material condition	−3.000
Bonus tolerance	.020
Plus the geometric tolerance	+ .014
Total tolerance	.034

The maximum material condition modifier allows the designer to capture all of the available tolerance.

Datums specified in order of precedence

When drawings are toleranced with the coordinate dimensioning system, datums are not specified. The lower and left edges on the drawing in Fig. 1-5 are *implied datums* because the holes are dimensioned from these edges. But which datum is more important, and is a third datum plane included in the datum reference frame? A rectangular part like this is usually placed in a datum reference frame consisting of three mutually perpendicular planes. When datums are not specified, machinists and inspectors are forced to make assumptions that could be very costly.

The parts placed in the datum reference frames in Fig. 1-6 show two interpretations of the drawing in Fig. 1-5. With the traditional method of tolerancing,

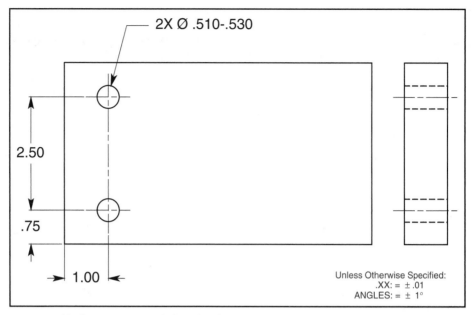

Figure 1-5 No datums are specified on this drawing.

it is not clear whether the lower edge of the part should be resting against the horizontal surface of the datum reference frame as in Fig. 1-6A or whether the left edge of the part should be in contact with the vertical surface of the datum reference frame as in Fig. 1-6B.

Manufactured parts are not perfect. It is clear that, when drawings are dimensioned with traditional tolerancing methods, a considerable amount of information is left to the machinists' and inspectors' judgment. If a part is to be inspected the same way every time, the drawing must specify how the part is to fit in the datum reference frame. All of the datums must be specified in order of precedence.

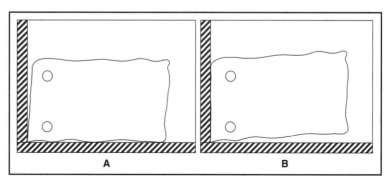

Figure 1-6 Possible datum interpretation.

Summary

- GD&T is a symbolic language used to specify the size, shape, form, orientation, and location of features on a part.
- GD&T was created to insure the proper assembly of mating parts, to improve quality, and to reduce cost.
- GD&T is a design tool.
- GD&T communicates design intent.
- This text is based on the standard *Dimensioning and Tolerancing ASME Y14.5M–1994*.
- The cylindrical tolerance zone defines a uniform distance from true position to the tolerance zone boundary.
- The maximum material condition symbol in the feature control frame is a modifier that allows a bonus tolerance.
- All of the datums must be specified in order of precedence.

Chapter Review

1. GD&T is a symbolic language used to specify the
   _____,_____,_____,_____
   and _____ of features on a part.
2. Features toleranced with GD&T reflect the _____
   between mating parts.
3. GD&T was designed to insure the assembly of _____
   _____, to improve _____ and to reduce _____ .
4. Geometric tolerancing allows the maximum available _____
   _____
   and consequently, the most _____ parts.
5. _____ is the current, authoritative reference
   document that specifies the proper application of GD&T.
6. Plus or minus tolerancing generates a _____ shaped
   tolerance zone.
7. _____ generates a cylindrical shaped tolerance zone to control an axis.
8. If the distance across a square tolerance zone is ± .005 or a total of .010,
   what is the approximate distance across the diagonal?_____.
9. Bonus tolerance equals the difference between the actual feature size and
   _____.
10. While processing, a rectangular part usually rests against a
    _____ consisting of three mutually
    perpendicular planes.

2

Dimensioning and Tolerancing Fundamentals

Many people know how to design parts and make drawings, yet they lack the basic knowledge to produce engineering drawings that conform to industry standards. Nonconforming drawings can be confusing, cause misunderstanding, and produce unacceptable parts. This chapter will familiarize the reader with some of the less well known but important standards based on dimensioning and tolerancing practices. All of the drawings in this book are dimensioned and toleranced with the inch system of measurement because most drawings produced in the United States are dimensioned with this system. Metric dimensioning is shown for illustration purposes only.

Chapter Objectives

After completing this chapter, you will be able to

- *Identify* fundamental drawing rules
- *Demonstrate* the proper way to specify units of measurement
- *Demonstrate* the proper way to specify dimensions and tolerances
- *Interpret* limits
- *Explain* the need for dimensioning and tolerancing on CAD/CAM database models

Fundamental Drawing Rules

Dimensioning and tolerancing shall clearly define engineering intent and shall conform to the following rules:

1. Each dimension shall have a tolerance except those dimensions specifically identified as reference, maximum, minimum, or stock.

2. Each feature shall be fully dimensioned and toleranced so that there is a complete description of the characteristics of each part. Use only the dimensions that are necessary for a full definition of the part. Reference dimensions should be kept to a minimum.

3. Each dimension shall be selected and arranged to satisfy the function and mating relationship of the part and shall not be subject to more than one interpretation.

4. The drawing should define the part without specifying a particular method of manufacturing.

5. A 90° angle applies where centerlines and lines representing features on a drawing are shown at right angles and no angle is specified.

6. A basic 90° angle applies where centerlines of features in a pattern or surfaces shown at right angles on a drawing are located or defined by basic dimensions and angles are not specified.

7. Unless otherwise specified, all dimensions are to be measured at 68°F (20°C). Measurements made at other temperatures may be adjusted mathematically.

8. All dimensions apply in the free-state condition except for nonrigid parts.

9. Unless otherwise specified, all geometric tolerances apply for the full depth, full length, and full width of the feature.

10. Dimensions and tolerances apply only at the drawing level where they are specified. For example, a dimension specified for a particular feature on a detailed drawing is not required for that feature on an assembly drawing.

Units of Linear Measurement

Units of linear measurement are typically expressed in either the inch system or the metric system. The system of measurement used on the drawing must be specified in a note, usually in the title block. A typical note reads: UNLESS OTHERWISE SPECIFIED, ALL DIMENSIONS ARE IN INCHES (or MILLIMETERS, as applicable). Some drawings have both the inch and the metric systems of measurement on them. On inch-dimensioned drawings where some dimensions are expressed in millimeters, the millimeter values are followed by the millimeter symbol, mm. On millimeter-dimensioned drawings where some dimensions are expressed in inches, the inch values are followed by the inch symbol, IN.

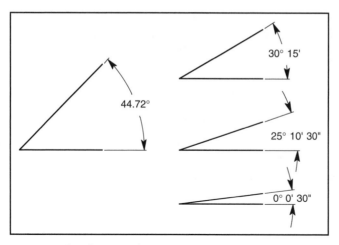

Figure 2-1 Angular measurement expressed with decimals and degrees, minutes, and seconds.

Units of Angular Measurement

Angular units of measurement are specified in either of two conventions as shown in Fig. 2.1.

- Degrees and decimal parts of a degree (44.72°)
- Degrees (°), minutes ('), and seconds (")

If degrees are assigned, the value is followed by the degree symbol (60°). If only minutes or seconds are indicated, the number of minutes or seconds shall be preceded by zero degrees (0°10′) or zero degrees and zero minutes (0°0′30″). Features appearing to be 90° on the drawing are, in fact, at an implied dimension of 90°. The tolerance for an implied 90° angle is the same as the tolerance for any other angle on the field of the drawing governed by a general note or the general, angular title block tolerance.

Two dimensions, 90° angles and zero dimensions, are not placed on the field of the drawing. A zero distance, such as the distance between two coaxial features, must be toleranced separately and cannot depend on the title block for its tolerance.

Types of Dimensions

There are two types of direct tolerancing methods:

- Limit dimensioning
- Plus and minus dimensioning

When using limit dimensioning, the high limit or the largest value is placed above the lower limit. If the tolerance is written on a single line, the lower limit precedes the higher limit separated by a dash. With plus and minus dimensioning, the dimension is followed by a plus or minus sign and the required tolerance.

TABLE 2-1 Inch and Millimeter Dimensions

	Decimal inch dimensions		Millimeter dimensions	
	Correct	Incorrect	Correct	Incorrect
1.	.25	0.25	0.25	.25
2.	4.500 ± .005	4.5 ± .005	4.5	4.500
3.			4	4.000

When specifying decimal inch dimensions on drawings (Table 2-1):

- A zero is *never* placed before the decimal point for values less than one inch. Some designers routinely place zeros before the decimal point for values less than one inch. This practice is incorrect and confusing for the reader.
- A dimension is specified with the same number of decimal places as its tolerance even if zeros need to be added to the right of the decimal point.

When specifying millimeter dimensions on drawings as described in Table 2-1:

- A zero *is* placed before the decimal point for values less than one millimeter.
- Zeros are *not* added to the right of the decimal point when dimensions are a whole number plus some decimal fraction of a millimeter. (This practice differs when tolerances are written bilaterally or as limits. See "Specifying Tolerances" below.)
- Neither a decimal point nor a zero is shown where the dimension is a whole number.

Specifying Linear Tolerances

When specifying decimal inch tolerances on drawings (Table 2-2):

- When a unilateral tolerance is specified and either the plus or the minus limit is zero, its zero value will have the same number of decimal places as the other limit and the appropriate plus or minus sign.
- Where bilateral tolerancing is specified, both the dimension and tolerance values have the same number of decimal places. Zeros are added when necessary.
- Where limit dimensioning and tolerancing is used, both values have the same number of decimal places even if zeros need to be added after the decimal place.

TABLE 2-2 Inch and Millimeter Tolerances

	Decimal inch tolerances		Millimeter tolerances	
	Correct	Incorrect	Correct	Incorrect
1.	+.000 .250 −.005	0 .250 −.005	0 40 −0.05	+.00 40 −.05
2.	+.025 .250 −.010	+.025 .25 −.010	+0.25 40 −0.10	+.25 40 −.1
3.	.500 .548	.5 .548	4.25 4.00	4.25 4

When specifying millimeter tolerances on drawings (Table 2-2):

- When a unilateral tolerance is specified and either the plus or the minus limit is zero, a single zero is shown and no plus or minus sign is used.
- Where bilateral tolerancing is specified, both tolerance values have the same number of decimal places. Zeros are added when necessary.
- Where limit dimensioning and tolerancing is used, both values have the same number of decimal places even if zeros need to be added after the decimal point.

Where basic inch dimensions are used, the basic dimension values are specified with the same number of decimal places as the associated tolerances as shown in Fig. 2-2. Where basic metric dimensions are used, the basic dimension values are specified with the practices shown in Table 2-1 for millimeter dimensoning.

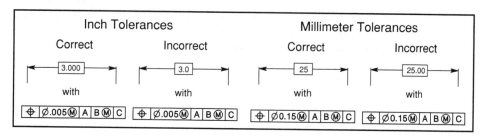

Figure 2-2 Basic dimensions and geometric tolerances have the same number of decimal places in the inch system. Basic millimeter dimensions conform to millimeter standards.

Specifying Angular Tolerances

When specifying angular tolerances in terms of degrees and decimal fractions of a degree on drawings as shown in Fig. 2-3, the angle and the plus and minus tolerance values are written with the same number of decimal places. When specifying angular tolerances in terms of degrees and minutes, the angle and

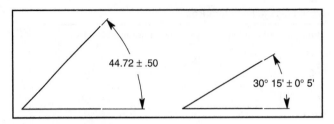

Figure 2-3 Angular tolerances.

the plus and minus tolerance values are written in degrees and minutes even if the number of degrees is zero.

Interpreting Dimensional Limits

All dimensional limits are absolute as shown in Table 2-3. Regardless of the number of decimal places, dimensional limits are used as if an infinite number of zeros followed the last digit after the decimal point.

TABLE 2-3   Dimensional Limits

4.0	Means	4.000...0
4.2	Means	4.200...0
4.25	Means	4.250...0

Dimensioning and Tolerancing for CAD/CAM Database Models

Many designers feel that solid model drawings produced with CAD/CAM programs do not need to be dimensioned or toleranced. The method of producing a design and transmitting that information to the manufacturing equipment is not the major cause of irregularity in parts. Although these systems may eliminate some human error, the major cause of part variation occurs as a result of a variety of other sources, such as

- Setup and stability of the part
- Quality and sharpness of tooling
- Quality and maintenance of machine tools
- Excessive clamping
- Size of the part
- The material the part is made from
- Heat treating
- Plating

None of these problems are addressed with the use of solid modeling programs. To quote *Dimensioning and Tolerancing ASME Y14.5M–1994*:

"CAUTION: If CAD/CAM database models are used and they do not include tolerances, then tolerance must be expressed outside of the database to reflect design requirements."

The most effective way to communicate design intent is through the proper use of geometric dimensioning and tolerancing.

Summary

- Units of linear measurement are typically expressed in either the inch system or the metric system and that system must be specified on the drawing.

- Angular units of measurement are specified either in degrees and decimal parts of a degree or in degrees, minutes, and seconds.

- There are two types of direct tolerancing methods, limit dimensioning and plus and minus dimensioning.

- A zero is never placed before the decimal point for values less than 1 inch. Even if zeros need be added to the right of the decimal point, dimensions are specified with the same number of decimal places as their tolerances.

- When a unilateral tolerance is specified and either the plus or the minus limit is zero, its zero value shall have the same number of decimal places as the other limit and the appropriate plus or minus sign. Where bilateral tolerancing is specified, both the dimension and tolerance values have the same number of decimal places.

- Where basic inch dimensions are used, the basic dimension values are written with the same number of decimal places as the associated tolerances.

- When specifying angular tolerances on drawings, the angle and the plus and minus tolerance values are expressed with the same number of decimal places.

- Regardless of the number of decimal places, dimensional limits are used as if an infinite number of zeros followed the last digit after the decimal point.

- If CAD/CAM database models do not include tolerances, they must be communicated outside of the database on a referenced document.

Chapter Review

1. Each dimension shall have a _____ except those dimensions specifically identified as reference, maximum, minimum, or stock.

2. Each feature shall be fully _____ and _____ so that there is a complete description of the characteristics of each part.

3. Each dimension shall not be subject to more than one _____.

4. The drawing should _____ the part without specifying a particular method of _____.

5. A _____ applies where centerlines and lines representing features on a drawing are shown at right angles and no angle is specified.

6. _____ applies where centerlines of features in a pattern or surfaces shown at right angles on a drawing are located or defined by basic dimensions and angles are not specified.

7. All dimensions are to be measured at _____ unless otherwise specified. Measurements made at other temperatures may be adjusted mathematically.

8. All dimensions apply in the _____ except for nonrigid parts.

9. All geometric tolerances apply for the _____ , _____ , and _____ of the feature unless otherwise specified.

10. Dimensions and tolerances apply only at the _____ where they are specified.

11. Units of linear measurement are typically expressed either in the _____ system or the _____ system.

12. Angular units of measurement are specified either in _____ or in _____ .

13. What two dimensions are not placed on the field of the drawing?

    _____

14. What are the two types of direct tolerancing methods?

    _____

15. For decimal inch tolerances, a _____ is never placed before the decimal point for values less than 1 inch.

16. For decimal inch tolerances, a dimension is specified with the same number of decimal places as its _____ .

17. For decimal inch tolerances, when a unilateral tolerance is specified and either the plus or minus limit is zero, its zero value will have _____ as the other limit and _____ .

18. For decimal inch tolerances, where bilateral tolerancing or limit dimensioning and tolerancing is used, both values have

    _____

19. Where basic dimensions are used, the basic dimension values are expressed with _____ .

20. Dimensional limits are used as if _____ followed the last digit after the decimal point.

21. If CAD/CAM database models are used and they do not include tolerances, then tolerance must be expressed _____ .

3

Symbols, Terms, and Rules

Symbols, terms, and rules are the basics of geometric dimensioning and tolerancing (GD&T). They are the alphabet, the definitions, and the syntax of this language. The GD&T practitioner must be very familiar with these symbols and know how to use them. It is best to commit them to memory. Can you imagine trying to read a book or write a composition without knowing the alphabet, without a good vocabulary, and without a working knowledge of how a sentence is constructed? A little memorization up front will save time and reduce frustration in the future.

Chapter Objectives

After completing this chapter, you will be able to

- *List* the 14 geometric characteristic symbols
- *Identify* the datum feature symbol
- *Explain* the elements of the feature control frame
- *List* the three material condition modifiers
- *Identify* the other symbols used with GD&T
- *Define* 12 critical terms
- *Explain* the four general rules.

Symbols

Geometric characteristic symbols

Geometric characteristic symbols are the essence of this graphic language. It is important not only to know each symbol but also to know how to apply these symbols on drawings. The 14 geometric characteristic symbols, shown in Fig. 3-1, are divided into five categories:

Pertainsto	Type of Tolerance	Geometric Characteristics	Symbol
Individual Feature Only	Form	STRAIGHTNESS	—
		FLATNESS	▱
		CIRCULARITY	○
		CYLINDRICITY	⌭
Individual Feature or Related Features	Profile	PROFILE OF A LINE	⌒
		PROFILE OF A SURFACE	⌓
Related Features	Orientation	ANGULARITY	∠
		PERPENDICULARITY	⊥
		PARALLELISM	//
	Location	POSITION	⌖
		CONCENTRICITY	◎
		SYMMETRY	≡
	Runout	CIRCULAR RUNOUT	↗
		TOTAL RUNOUT	↗↗

Figure 3-1   Geometric characteristic symbols.

- Form
- Profile
- Orientation
- Runout
- Location

It is important to learn these symbols in their respective categories because many characteristics that apply to one geometric control also apply to other geometric controls in the same category. For example, datums are not appropriate for any of the form controls. Notice that form controls pertain only to individual features. In other words, form controls are not related to datums. Orientation, location, and runout controls must have datums since they are related features. Profile controls may have datums, or not, as required.

The datum feature symbol

The datum feature symbol consists of a capital letter enclosed in a square box. It is connected to a leader directed to the datum ending in a triangle. The triangle may be solid or open. The datum identifying letters may be any letter of the alphabet except I, O, and Q. Multiple letters such as AA through AZ, BA through BZ, etc., may be used if a large number of datums are required. The datum feature symbol is used to identify physical features of a part. The datum feature symbol must **not** be attached to centerlines, center planes, or axes. It may be directed to the outline or extension line of a feature such as datums A through G shown in the top two drawings of Fig. 3-2. The datum feature symbol may also be associated with a leader or dimension line as shown in the lower two figures. If only a leader is used, the datum feature symbol is attached to the tail, such as datum J in Fig. 3-2. A datum feature symbol is typically attached to a feature control frame directed to the datum with a leader, such as datums K, M, and N. If the datum feature symbol is placed in line with a dimension line or on a feature control frame associated with a size feature, the size feature is the datum. For example, in Fig. 3-2, datum R is the 3.00-inch size feature between the top and bottom surfaces, and datum S is the 1.00-inch slot.

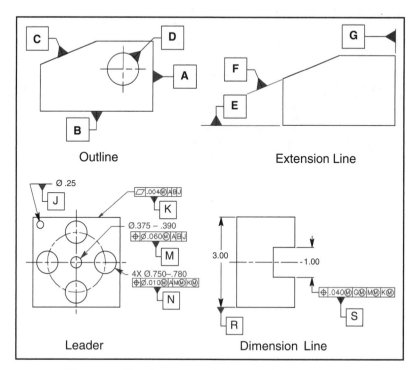

Figure 3-2 Four views illustrating methods of attaching datum feature symbols to features.

The feature control frame

The feature control frame in the GD&T language is like a sentence in the English language—it is a complete tolerancing thought. All of the geometric tolerancing for a feature, or pattern of features, is contained in one or more feature control frames. Just as in any other language, the feature control frame must be properly and completely written.

One of the fourteen geometric characteristic symbols always appears in the first compartment of the feature control frame. The second compartment is the tolerance section. In this compartment, there is, of course, the tolerance followed by any appropriate modifiers. Figure 3-3 shows a feature control frame with the maximum material condition (MMC) modifier (circle M). The tolerance is preceded by a diameter symbol if the tolerance zone is cylindrical. If the tolerance zone is not cylindrical, then nothing precedes the tolerance. The final section is reserved for datums and any appropriate material condition modifiers. If the datum is a size feature, then a material condition applies; if no material condition modifier is specified, then "regardless of feature size" (RFS) automatically applies. Datums are arranged in the order of precedence or importance. The first datum to appear in the feature control frame, the primary datum, is the most important datum. The second datum, the secondary datum, is the next most important datum, and the tertiary datum is the least important. Datums do not have to be specified in alphabetical order.

The feature control frame in Fig. 3-4 may be read as follows. The axis of the hole must be positioned within a cylindrical tolerance zone of .014 in diameter

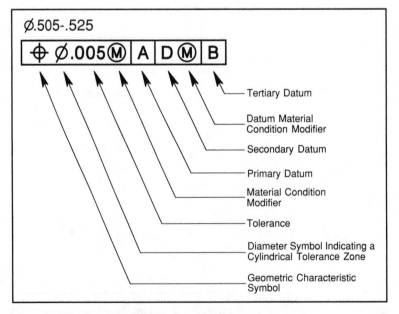

Figure 3-3 The feature control frame explained.

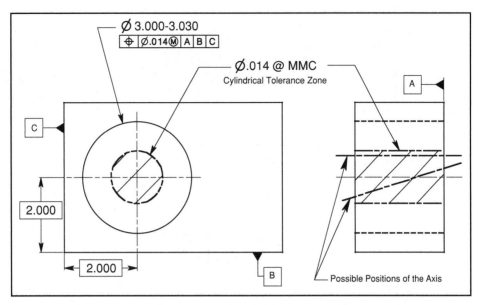

Figure 3-4 A feature control frame specifying the tolerance zone size, shape, and relationship to its datums.

at MMC (circle M). The tolerance zone is perpendicular to datum A, located up from datum B and over from datum C. If the hole is produced at its MMC, Ø3.000, the diameter of the tolerance zone is .014. If it is produced at Ø3.020, as shown in Chapter 1, the diameter of the tolerance zone is .034.

Feature control frames may be attached to features with extension lines, dimension lines, or leaders. For flat surfaces, a side or end of a feature control frame may be attached to an extension line as shown in Fig. 3-5A. Even a corner of the feature control frame may be attached to an extension line extending from a surface at an angle to the horizontal plane. A feature control frame may be placed beneath a dimension or attached to an extension of a dimension line as in Fig. 3-5B. Finally, a feature control frame may be attached to a leader directed to a feature surface or placed beneath a dimension directed with a leader to a size feature such as a hole shown in Fig. 3-5C.

The composite feature control frame consists of one geometric characteristic symbol followed by two tolerance and datum sections as shown in Fig. 3-6A. The lower segment is a refinement of the upper segment. The two single-segment feature control frames (Fig. 3-6B) consist of two complete feature control frames, one above the other, with different datum references as shown. The lower segment is a refinement of the upper segment. In Fig. 3-6C, a single feature control frame may have one or more feature control frames refining the tolerance of specific feature characteristics. These controls will be discussed in more detail in later chapters.

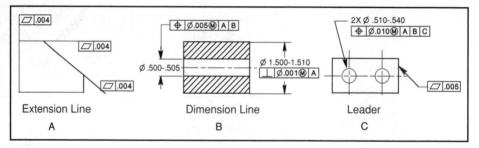

Figure 3-5 Feature control frames attached to features.

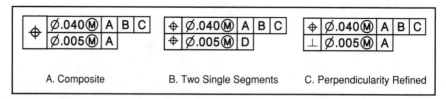

Figure 3-6 Composite, two single segments, and refined feature control frames.

Material conditions

One of the major advantages of using GD&T is the ability to specify how a particular size feature applies. If a size feature has a geometric tolerance or if it is used as a datum, then the tolerance or datum applies at one of these material conditions (Table 3-1):

- Regardless of feature size
- Maximum material condition
- Least material condition

In previous revisions of the geometric tolerancing standard, the symbol for RFS was a circle S. This symbol is no longer used because RFS, in the current standard, is the default material condition modifier. If no material condition symbol is specified for the tolerance or datum reference, the feature automatically applies at RFS, which means that the tolerance is the same, no matter what size the feature has been produced within its limits of size. A tolerance specified at RFS is only the tolerance specified in the feature control frame, and no bonus tolerance is added. Geometric tolerances specified at RFS are often used when tolerancing high speed, rotating parts, or when symmetrical relationships are required. Material condition modifiers are explained in more detail in Chapter 7.

TABLE 3-1 Material Condition Symbols

Material condition modifier	Abbreviation	Symbol
Regardless of feature size	RFS	None
Maximum material condition	MMC	Ⓜ
Least material condition	LMC	Ⓛ

Where the **Maximum Material Condition Modifier (circle M)** is specified to modify a size feature in a feature control frame, the following two requirements apply:

- The specified tolerance applies at the **Maximum Material Condition (MMC)** size of a feature. The MMC size of a feature is the largest shaft and the smallest hole. The MMC modifier (circle M) is not to be confused with the MMC size of a feature as shown in Fig. 3-7.
- As the size of the feature departs from MMC toward LMC, a bonus tolerance is gained in the exact amount of such departure. Bonus tolerance is the difference between the actual feature size and the MMC of the feature. The bonus tolerance is added to the geometric tolerance specified in the feature control frame. The MMC is the most common of the material conditions. It is often used to tolerance parts that fit together in a static assembly, for example, an assembly that is bolted together.

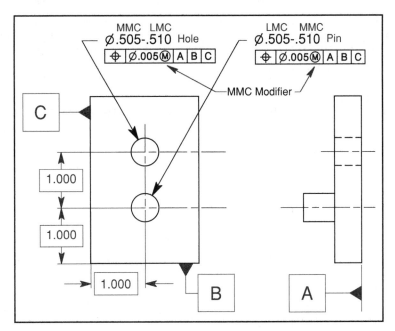

Figure 3-7 Hole and pin drawing for bonus calculation at MMC.

The following formulas are used to calculate the bonus tolerance and total positional tolerance at MMC (Table 3-2):

- **Bonus** equals the difference between the **Actual Feature Size** and **MMC.**
- **Bonus** plus **Geometric Tolerance** equals **Total Positional Tolerance.**

TABLE 3-2 The Increase in Bonus and Total Tolerance as the Size of the Features Departs from MMC Toward LMC

Actual feature size	MMC	Bonus	Geometric tolerance	Total positional tolerance
Internal Feature (Hole)				
MMC .505	.505	.000	.005	.005
.506	.505	.001	.005	.006
.507	.505	.002	.005	.007
.508	.505	.003	.005	.008
.509	.505	.004	.005	.009
LMC .510	.505	.005	.005	.010
External Feature (Pin)				
MMC .510	.510	.000	.005	.005
.509	.510	.001	.005	.006
.508	.510	.002	.005	.007
.507	.510	.003	.005	.008
.506	.510	.004	.005	.009
LMC .505	.510	.005	.005	.010

Where the **Least Material Condition Modifier (circle L)** is specified to modify a size feature in a feature control frame, the following two requirements apply:

- The specified tolerance applies at the **LMC** size of a feature. The LMC size of a feature is the smallest shaft and the largest hole. The LMC modifier (circle L) is not to be confused with the LMC size of a feature.
- As the size of the feature departs from LMC toward MMC, a bonus tolerance is gained in the exact amount of such departure. Bonus tolerance is the difference between the actual feature size and the LMC of the feature. The bonus tolerance is added to the geometric tolerance specified in the feature control frame. LMC is used to maintain a minimum distance between features. The LMC is seldom used. Functional gages cannot be used to inspect features specified at LMC.

Other symbols used with geometric tolerancing

A number of other symbols used with GD&T are listed in Fig. 3-8. They are discussed in more detail below and in subsequent chapters.

All Around	⌲	Free State	Ⓕ
Between	↔	Projected Tolerance Zone	Ⓟ
Number of Places	X	Tangent Plane	Ⓣ
Counterbore/Spotface	⌴	Radius	R
Contersink	∨	Radius, Controlled	CR
Depth/Deep	↧	Spherical Radius	SR
Diameter	⌀	Spherical Diameter	S⌀
Dimension, Basic	1.000	Square	□
Dimension, Reference	(60)	Statistical Tolerance	⟨ST⟩
Dimension Origin	◄—⬥	Datum Target	⌀.500 / A1
Arc Length	⌒110	Target Point	✕
Conical Taper	⊳	Slope	◿

Figure 3-8 Other symbols used on prints.

The **All Around** and the **Between** symbols are used with the profile control as shown in Fig. 3-9. When a small circle is placed at the joint of the leader, a profile tolerance is specified all around the surface of the part. The between symbol in the drawing above indicates that the tolerance applies between points X and Z on the portion of the profile where the leader is pointing.

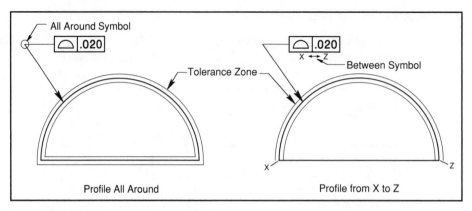

Figure 3-9   All around and between symbols.

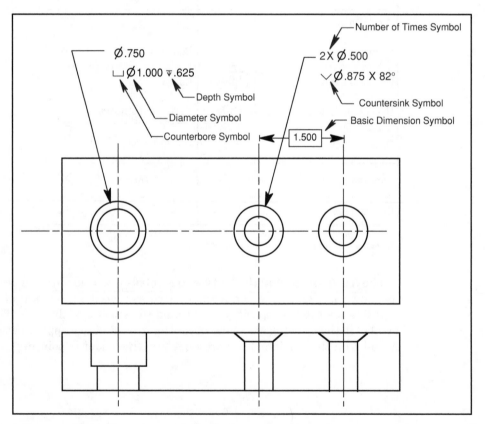

Figure 3-10   Counterbore, countersink, depth, diameter, and basic dimension symbols.

The **Counterbore** and **Countersink** symbols are indicated as shown in Fig. 3-10. The counterbore symbol is also used to indicate a **Spotface** operation. The **Depth** symbol is used to indicate the depth of a feature. The **Basic Dimension** has a box around the dimension. The title block tolerance does not apply to basic dimensions. The tolerance associated with a basic dimension usually appears in a feature control frame or a note.

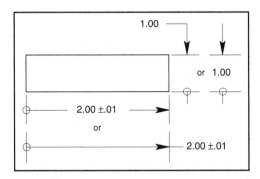

Figure 3-11 Dimension origin symbol.

The **Dimension Origin** symbol indicates that the measurement of a feature starts at the origin, which is the end of the dimension line that has the circle. Fig. 3-11 shows several ways to specify the dimension origin symbol.

A **Radius** is a straight line connecting the center and the periphery of a circle or sphere.

The **Radius symbol R**, shown in Fig. 3-12, defines a tolerance zone bounded by a maximum radius arc and a minimum radius arc that are tangent to the adjacent surfaces. The surface of the toleranced radius must lie within this tolerance zone.

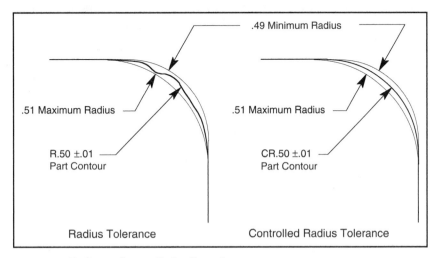

Figure 3-12 Radius and controlled radius tolerances.

The **Controlled Radius symbol CR** also defines a tolerance zone bounded by a maximum radius arc and a minimum radius arc that are tangent to the adjacent surfaces. However, the surface of the controlled radius must not only lie within this tolerance zone but also be a fair (smooth) curve with no reversals. In addition, at no point on the radius can the curve be greater than the maximum limit, nor smaller than the minimum limit. Additional requirements may be specified in a note.

The **Spherical Radius SR** and **Spherical Diameter SØ** symbols, shown in Fig. 3-8, indicate the radius and the diameter of a sphere.

The free state symbol specifies that tolerances for nonrigid features, subject to free state variation, apply in their "free state."

The projected tolerance zone symbol specifies that the tolerance zone is to be projected into the mating part.

The tangent plane symbol specifies that if a precision plane contacting the high points of a surface falls within the specified tolerance zone, the surface is in tolerance.

The **Statistical Tolerance** symbol indicates that the tolerance is based on a statistical tolerance. The statistical tolerance symbol may also be applied to a size tolerance. The four modifiers mentioned above are placed in the feature control frame after the tolerance and any material condition symbols as shown in Fig. 3-13.

The **Square** symbol preceding a dimension specifies that the toleranced feature is square and the dimension applies in both directions as shown in Fig. 3-14. The square symbol applies to square features the way a diameter symbol applies to cylindrical features.

Conical Taper is defined as the ratio of the difference between two diameters, perpendicular to the axis of a cone, divided by the length between the two diameters.

$$\text{Taper} = (D - d)/L$$

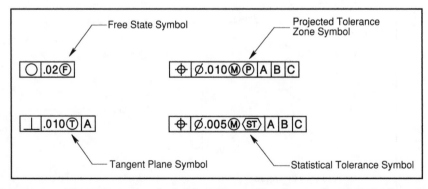

Figure 3-13 Free state, projected tolerance zone, tangent plane, and statistical tolerance symbols.

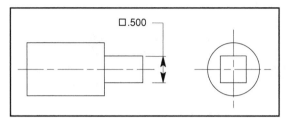

Figure 3-14 Square symbol.

Here, D is the larger diameter, d is the smaller diameter, and L is the length between the two diameters.

Slope is defined as the ratio of the difference in heights at both ends of an inclined surface, measured at right angles above a base line, and divided by the length between the two heights.

$$\text{Slope} = (H - h)/L$$

Here, H is the larger height, h is the smaller height, and L is the length between the two heights.

A **Reference Dimension** is a numerical value without a tolerance, used only for general information. It is additional information and may not be used for manufacturing or inspection. The reference dimension is indicated by placing parenthesis around the numerical value as shown in Fig. 3-15.

The **Arc Length symbol** shown in Fig. 3-8 indicates that a linear dimension is used to measure an arc along its curved outline.

Datum Target symbols and **Datum Target Points** are explained in Chapter 4, Datums.

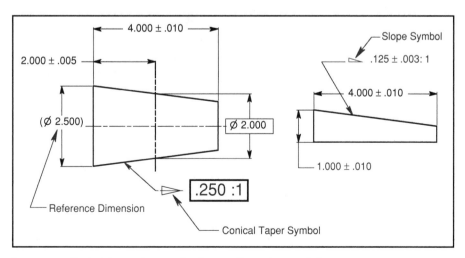

Figure 3-15 Conical taper, slope, and reference dimension symbols.

Terms

The names and definitions of many GD&T terms have very specific meanings. In some cases they are quite different from general English usage. To be able to function in this language, it is important for each GD&T practitioner to be very familiar with these 12 terms.

1. *Actual mating envelope:* The actual mating envelope is defined separately for internal and external features.
   - *External feature*: The actual mating envelope for an external feature of size is the smallest, similar, perfect, feature counterpart that can be circumscribed around the feature so that it just contacts the surface(s) at the highest points. For example, the actual mating envelope of a pin is the smallest precision sleeve that just fits over the pin contacting the surface at the highest points.
   - *Internal feature*: The actual mating envelope for an internal feature of size is the largest, similar, perfect, feature counterpart that can be inscribed within the feature so that it just contacts the surface(s) at the highest points. For example, the actual mating envelope of a hole is the largest precision pin that just fits inside the hole contacting the surface at the highest points.

   The actual mating envelope of a feature, controlled by an orientation or a position tolerance, is oriented to the specified datum(s). For example, the actual mating envelope may be the largest pin that fits through the hole and is perpendicular to the primary datum plane illustrated in Fig. 3-16.

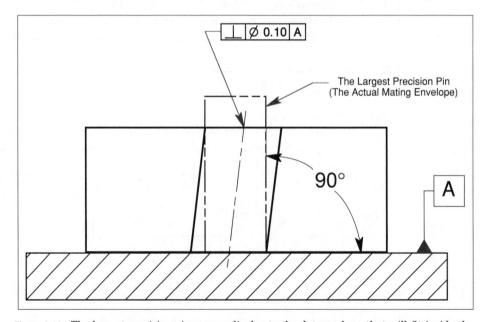

Figure 3-16 The largest precision pin, perpendicular to the datum plane that will fit inside the hole.

2. *Basic dimension:* A basic dimension is a numerical value used to describe the theoretically exact size, profile, orientation, or location of a feature or datum target. Basic dimensions are used to define or position tolerance zones. Title block tolerances do not apply to basic dimensions. The tolerance associated with a basic dimension usually appears in a feature control frame or a note.

3. *Datum:* A datum is a theoretically exact point, line, or plane derived from the true geometric counterpart of a specified datum feature. A datum is the origin from which the location or geometric characteristics of features of a part are established.

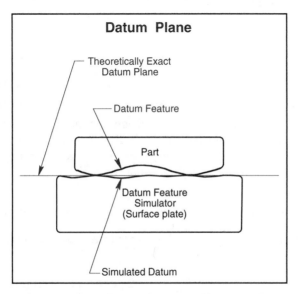

Figure 3-17 The difference between a datum, a datum feature, and a datum feature simulator.

4. *Datum feature:* A datum feature is an actual feature on a part used to establish a datum.

5. *Datum feature simulator:* A datum feature simulator is a real surface with a sufficiently precise form, such as a surface plate, machine table, or gage pin used to contact datum features to establish simulated datums. The datum is understood to exist in and be simulated by the datum feature simulator (Fig. 3-17).

6. *Feature:* A feature is a physical portion of a part, such as a flat surface, pin, hole, tab, or slot.

7. *Feature of size* (also **Size Feature** and **Feature Subject to Size Variations**): Features of size are features that have a size dimension. A feature of size takes four forms:

- Cylindrical surfaces
- Two opposed parallel surfaces
- A spherical surface
- Two opposed elements

Cylindrical surfaces and two opposed parallel surfaces are the most common features of size.

8. *Least material condition (LMC):* The least material condition of a feature of size is the least amount of material within the stated limits of size. For example, the minimum shaft diameter or the maximum hole diameter.

9. *Maximum material condition (MMC):* The maximum material condition of a feature of size is the maximum amount of material within the stated limits of size, for example, the maximum shaft diameter or the minimum hole diameter.

10. *Regardless of feature size (RFS):* Regardless of feature size is a material condition modifier used in a feature control frame to indicate that a geometric tolerance or datum reference applies at each increment of size of the feature within its limits of size. RFS specifies that no bonus tolerance is allowed.

11. *Resultant condition:* The resultant condition of a feature specified at MMC is a variable boundary generated by the collective effects of the LMC limit of size of a feature, the specified geometric tolerance, and any applicable bonus tolerance. Features specified with an LMC modifier also have a resultant condition.

Extreme resultant condition calculations for features toleranced at MMC:

External Features (Pin)	**Internal Features (Hole)**
LMC	LMC
Minus Geometric Tolerance @ MMC	*Plus* Geometric Tolerance @ MMC
Minus Applicable Bonus Tolerance	*Plus* Applicable Bonus Tolerance
Resultant Condition	Resultant Condition

12. *True position:* True position is the theoretically exact location of a feature established by basic dimensions. Tolerance zones are located at true position.

13. *Virtual condition:* The virtual condition of a feature specified at MMC is a constant boundary generated by the collective effects of the MMC limit of size of a feature and the specified geometric tolerance. Features specified with an LMC modifier also have a virtual condition.

Virtual condition calculations:

External Features (Pin)	**Internal Features (Hole)**
MMC	MMC
Plus Geometric Tolerance @ MMC	*Minus* Geometric Tolerance @ MMC
Virtual Condition	Virtual Condition

14. *Worst-case boundary:* The worst-case boundary of a feature is a general term that describes the smallest or largest boundary (i.e., a locus) generated by the collective effects of the MMC or LMC of the feature and any applicable geometric tolerance.
- Inner boundary specified at MMC
  The worst-case inner boundary is the virtual condition of an internal feature and the extreme resultant condition of an external feature.
- Outer boundary specified at MMC
  The worst-case outer boundary is the extreme resultant condition of an internal feature and the virtual condition of an external feature.

Features specified with an LMC modifier also have worst-case boundaries.

Rules

There are four rules that apply to drawings in general, and to GD&T in particular. They govern specific relationships of features on a drawing. It is important for each GD&T practitioner to know these rules and to know how to apply them.

Rule #1

Rule #1 states that where only a tolerance of size is specified, the limits of size of an individual feature of size prescribe the extent to which variations in its **geometric form**, as well as its **size**, are allowed. No element of a feature shall extend beyond the MMC boundary of perfect form. The form tolerance increases as the actual size of the feature departs from MMC toward LMC. There is no perfect form boundary requirement at LMC.

In Fig. 3-18, the MMC of the pin is 1.020. The pin may, in no way, fall outside this MMC boundary or envelope of perfect form. That is, if the pin is produced at a diameter of 1.020 at each and every cross section, it must not be bowed or out of circularity in any way. If the pin is produced at a diameter of 1.010 at each and every cross section, it may be out of straightness and/or out of circularity by a total of .010. If the pin is produced at a diameter of 1.000, its LMC, it may vary from perfect form the full .020 tolerance.

Rule #1 does not apply to stock or to features subject to free state variation in the unrestrained condition. When the word *stock* is specified on a drawing, it indicates bar, plate, sheet, etc., as it comes from the supplier. Stock items are manufactured to industry or government standards and are not controlled by Rule #1. Stock is used as is, unless otherwise specified by a geometric tolerance or note. Rule #1 does not apply to parts that are flexible and are to be measured in their **free state**.

Perfect form at MMC is not required if it is desired to allow the surface(s) of a feature to exceed the boundary of perfect form at MMC. In such cases, the note, PERFECT FORM AT MMC NOT REQD, may be specified on the drawing.

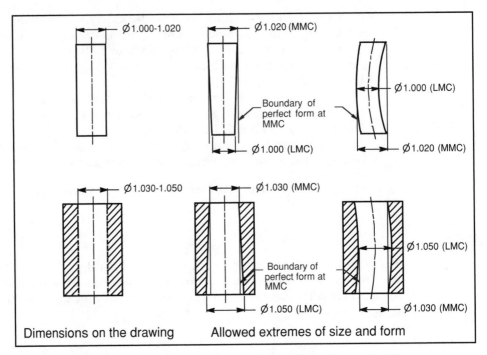

Figure 3-18 Rule #1 – examples of size and form variations allowed by the size tolerance.

The relationship between individual features is **not** controlled by the limits of size. If features on a drawing are shown coaxial, or symmetrical to each other and are not controlled for location, the drawing is incomplete. Figure 3-19A is incomplete because there is no control of coaxiality between the inside diameter and the outside diameter. Figure 3-19B shows one way of specifying the coaxiality of the inside and outside diameters.

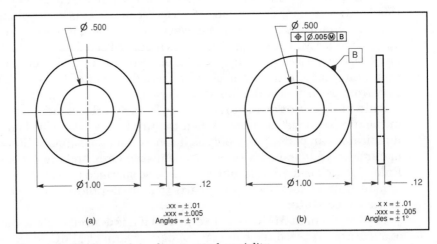

Figure 3-19 The limits of size do not control coaxiality.

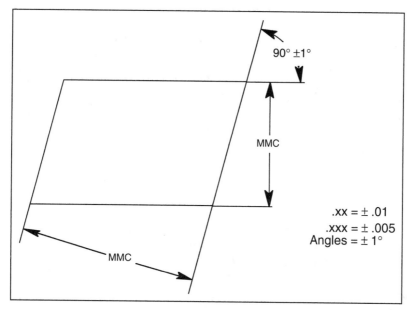

Figure 3-20   Angularity tolerance controls the angularity between individual features.

As shown by the part in Fig. 3-20, the perpendicularity between size features is not controlled by the size tolerance. There is a misconception that the corners of a rectangle are perfectly square if the sides are produced at MMC. If no orientation tolerance is specified, perpendicularity is controlled, not by the size tolerance, but by the angularity tolerance. The right angles of the rectangle in Fig. 3-20 may fall between 89° and 91° as specified by the angular tolerance in the title block.

Rule #2

Rule #2 states that **RFS** automatically applies, in a feature control frame, to individual tolerances of size features and to datum features of size. **MMC** and **LMC** must be specified when these conditions are required.

In Fig. 3-21, both the feature being controlled and the datum are size features. The feature control frame labeled **A** has no modifiers. Therefore, the coaxiality tolerance and the datum, controlled by the feature control frame labeled **A**, apply at RFS. The feature control frame labeled **B** has an MMC modifier (circle M) following the tolerance and datum D. If the Ø2.000 feature is controlled by the feature control frame labeled **B**, both the tolerance and the datum apply at MMC, and additional tolerance is allowed as the features depart from MMC toward LMC.

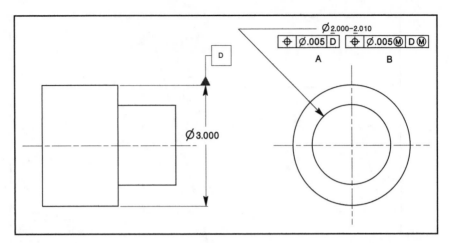

Figure 3-21   Feature control frames specified with RFS and MMC.

The pitch diameter rule

Each tolerance of orientation or position and datum reference specified for screw threads applies to the axis of the thread derived from the pitch diameter. Exceptions to this rule may be specified by placing a note, such as MAJOR DIA or MINOR DIA, beneath the feature control frame, or beneath or adjacent to, the datum feature symbol.

Each tolerance of orientation or position and datum reference specified for gears and splines must designate the specific feature, such as MAJOR DIA, PITCH DIA, or MINOR DIA, at which each applies. The note is placed beneath the feature control frame, or beneath or adjacent to, the datum feature symbol.

The virtual condition rule

Where a **datum feature of size** is controlled by a **geometric tolerance** and that datum is specified as a **secondary or tertiary datum**, the datum applies at virtual condition with respect to **orientation**.

In Fig. 3-22, the center hole

- Is a **datum**, datum D;
- Is a **size feature**;
- Has a **geometric tolerance**, and in fact, this hole has two geometric tolerances: position and perpendicularity
- Is specified as a **secondary datum** in the feature control frame controlling the four-hole pattern.

Since the conditions for the virtual condition rule exist, datum D applies at virtual condition. But datum D has two geometric controls, which means

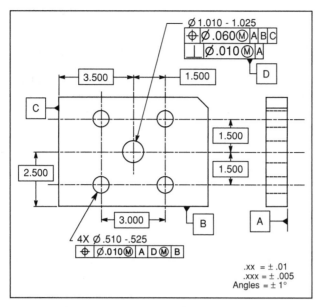

Figure 3-22 The center hole, datum D, applies at virtual condition with respect to orientation.

it has two virtual conditions. The first is the virtual condition for the position tolerance controlling the **location** of the center hole to datums B and C, and the second is the virtual condition for the perpendicularity control to refine the **orientation** tolerance of the center hole perpendicular to datum A.

The question is, which virtual condition should be used to calculate the shift tolerance? Shift tolerance is the additional tolerance gained when the **datum feature** departs from MMC or virtual condition toward LMC. Datum D applies at virtual condition with respect to perpendicularity because the relationship between datum plane A and datum axis D is orientation, not location.

If a gage is used to inspect this part, the primary datum on the part (datum feature A) must rest with a minimum of three points of contact against datum surface A on the gage. If the hole is out of perpendicularity with respect to datum A, the gage pin must be made at virtual condition, or it will not fit through the hole at its worst-case condition.

The virtual condition calculation for datum D is:

MMC	1.010
Minus the geometric tolerance (Perpendicularity)	− .010
Virtual condition (Orientation)	1.000

If the hole, datum D, had only a position control, we would use the position tolerance to calculate the virtual condition since the position control is a compound control that locates and orients size features simultaneously to the same tolerance.

If datum D is actually produced a Ø1.020 and the virtual condition is a Ø1.000, then the four-hole pattern can shift in any direction within a cylindrical tolerance zone of .020 in diameter. The virtual condition rule and shift tolerance will be discussed in more detail in later chapters.

Summary

- There are **14 geometric characteristic symbols**. They are divided into five categories: form, profile, orientation, runout, and location.

- The **datum feature symbol** consists of a capital letter enclosed in a square box. It is connected to a leader directed to the datum ending in a triangle.

- The datum feature symbol is used to identify physical features of a part. It must not be attached to centerlines, center planes, or axes.

- Datum feature symbols placed in line with a dimension line or on a feature control frame associated with a size feature identify the size feature as the datum.

- The feature control frame is the sentence of the GD&T language.

- Feature control frames may be attached to features with extension lines, dimension lines, and leaders.

- The composite feature control frame consists of one geometric characteristic symbol followed by two tolerance and datum sections.

- If no material condition symbol is specified for the tolerance or datum reference of a size feature in a feature control frame, RFS automatically applies.

- An RFS tolerance is only the tolerance specified in the feature control frame; no bonus tolerance is added.

- Where the MMC **symbol** is specified, the tolerance applies at the MMC, and applicable bonus tolerances are added to the geometric tolerance.

- MMC is the most common of the material conditions and is often used when parts are to be joined in a static assembly.

- Where the LMC **symbol** is specified, the tolerance applies at the LMC, and applicable bonus tolerances are added to the geometric tolerance.

- LMC is used to maintain a minimum distance between features.

- A number of other symbols used with GD&T are listed in Fig. 3-8. The reader should be able to recognize each of these symbols.

- The names and definitions of many GD&T concepts have very specific meanings. To be able to properly read and apply GD&T, it is important to be very familiar with these 12 terms.

- There are four rules that apply to drawings. It is important to know these rules and how to apply them.

- **Rule #1** states that where only a tolerance of size is specified, the limits of size of an individual feature prescribe the extent to which variations in its **geometric form**, as well as **size**, are allowed.

- **Rule #2** states that, in a feature control frame, RFS automatically applies to individual tolerances of size and to datum features of size. MMC and LMC must be specified when these conditions are required.

- **The Pitch Diameter Rule** states that each geometric tolerance or datum reference specified for screw threads applies to the axis of the thread derived from the pitch diameter.

- **The Virtual Condition Rule** states that where a datum feature of size is controlled by a geometric tolerance and is specified as a secondary or tertiary datum, the datum applies at virtual condition with respect to orientation.

Chapter Review

1. The second compartment of the feature control frame is the _____ compartment.

2. What type of geometric controls has no datums? _____

3. Which of the location controls is the most common? _____

4. What type of geometric controls indicates an angular relationship with specified datums? _____

5. What is the name of the symbol that must identify physical features of a part and shall not be applied to centerlines, center planes, or axes? _____

6. Datum identifying letters may be any letter of the alphabet except what letters? _____

7. If the datum feature symbol is placed in line with a dimension line or on a feature control frame associated with a size feature, then the datum is what? _____

8. One of the fourteen geometric characteristic symbols always appears in the _____ compartment of the feature control frame.

Pertainsto	Type of Tolerance	Geometric Characteristics	Symbol
Individual Feature Only	Form	_____ _____ _____ _____	
Individual Feature or Related Features	Profile	_____ _____	
Related Features	Orientation	_____ _____ _____	
	Location	_____ _____ _____	
	Runout	_____ _____	

Figure 3-23 Geometric characteristic symbols.

9. Write the names and geometric characteristic symbols where indicated in Fig. 3-23.

10. The tolerance is preceded by a diameter symbol only if the tolerance zone is _____.

11. Datums are arranged in the order of _____.

12. Write the name, abbreviation, and symbol for the three material condition modifiers.

_____   _____   _____

_____   _____   _____

_____   _____   _____

13. Which modifier specifies that the tolerance is the same, no matter what size the feature is within its size limits? _____.

14. The MMC modifier specifies that the tolerance applies at the

_____   _____   _____ of the feature.

15. The MMC modifier specifies that as the actual size of the feature departs from MMC toward LMC, a _____ is achieved in the exact amount of such departure.

16. The bonus tolerance equals the difference between the _____.

_____

17. The total positional tolerance equals the sum of the _____ tolerance and the _____ tolerance.

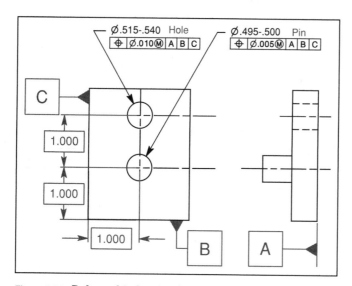

Figure 3-24 Refer to this drawing for questions 18 through 25.

	Hole	Pin
18. What is the MMC?	_____	_____
19. What is the LMC?	_____	_____
20. What is the geometric tolerance?	_____	_____
21. What material condition modifier is specified? _____		

22. What datum(s) control(s) perpendicularity? _____

23. What datum(s) control(s) location? _____

24. Complete the table below.

TABLE 3-3   Bonus Tolerance for Holes

| Actual feature size | Internal feature (Hole) | | | Total positional tolerance |
	MMC	Bonus	Geometric tolerance	
MMC 0.515				
0.520				
0.525				
0.530				
0.535				
LMC 0.540				

25. Complete the table below.

TABLE 3-4   Bonus Tolerance for Pins

| Actual feature size | External feature (Pin) | | | Total positional tolerance |
	MMC	Bonus	Geometric tolerance	
MMC 0.500				
0.499				
0.498				
0.497				
0.496				
LMC 0.495				

Using the drawing in Fig. 3-24, complete Tables 3-3 and 3-4 above.

26. The all around and between symbols are used with what control?
_____

27. What is the name of an actual feature on a part used to establish a datum?
_____

28. A numerical value used to specify the theoretically exact size, profile, orientation, or location of a feature is called? _____

29. What is the theoretically exact point, line, or plane derived from the true geometric counterpart of a specified datum feature called? _____

30. What is a real surface with a sufficiently precise form, such as a surface plate or machine table, used to contact datum features to establish simulated datums called? _____

Name	Symbol	Name	Symbol
All Around		Free State	
Between		Projected Tolerance Zone	
Number of Places		Tangent Plane	
Counterbore/Spotface		Radius	
Countersink		Radius, Controlled	
Depth/Deep		Spherical Radius	
Diameter		Spherical Diameter	
Dimension, Basic	1.000	Square	
Dimension, Reference	60	Statistical Tolerance	
Dimension Origin		Datum Target	
Arc Length	110	Target Point	
Conical Taper		Slope	

Figure 3-25 Geometric tolerancing symbols.

31. Draw the indicated geometric tolerancing symbols in the spaces on Fig. 3-25.

32. What is the name of a physical portion of a part, such as a surface, pin, hole, tab, or slot? _____

33. What is the name of a feature that has a dimension such as a cylindrical surface or two opposed parallel surfaces? _____

34. What kind of features always apply at MMC, LMC, or RFS? _____

35. What is the maximum amount of material within the stated limits of size of a size feature called? _____

36. What is a feature of size with the least amount of material within the stated limits of size called? _____

37. What is the term used to indicate that a specified geometric tolerance or datum reference applies at each increment of size of a feature within its limits of size? _____

38. What is the theoretically exact location of a feature established by basic dimensions called? _____

39. What is a constant boundary generated by the collective effects of the MMC limit of size of a feature and the applicable geometric tolerance called? _____

40. Where only a tolerance of size is specified, the limits of size of an individual feature prescribe the extent to which variations in its **geometric form,** as well as **size**, are allowed. This statement is the essence of _____

41. The form tolerance increases as the actual size of the feature departs from _____ toward _____.

42. If features on a drawing are shown coaxial, or symmetrical to each other and not controlled for _____ , the drawing is incomplete.

43. If there is no orientation control specified for a rectangle on a drawing, the perpendicularity is controlled, not by the _____ , but by the _____ tolerance.

44. Rule #2 states that _____ automatically applies, to individual tolerances of size features and to datum features of size.

45. Each geometric tolerances or datum reference specified for screw threads applies to the axis of the thread derived from the _____.

46. Each geometric tolerance or datum reference specified for gears and splines must designate the specific feature at which each applies such as _____.

47. Where a datum feature of size is controlled by a geometric tolerance and is specified as a secondary or tertiary datum, the datum applies at _____ with respect to orientation.

Problems

Figure 3-26   Material condition symbols – Problem 1.

1. Read the complete tolerance in each feature control frame in Fig. 3-26, and write them below (Datum A is a size feature).
   A. _____

   _____

   B. _____

   _____

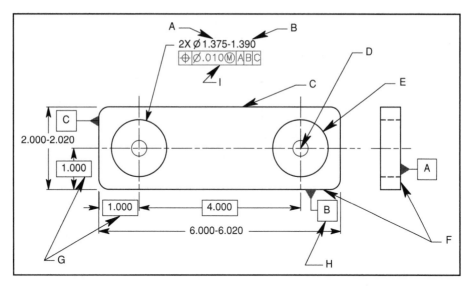

Figure 3-27 Definitions – Problem 2.

2. Place the letters of the items on the drawing in Fig. 3-27 next to the terms below. Make a dash next to the terms not shown.

_____ Datum _____ Basic Dimension _____ Feature control frame

_____ MMC _____ Feature _____ True Position

_____ LMC _____ Feature of Size _____ Datum feature symbol

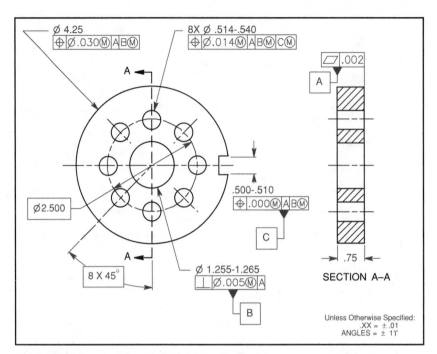

Figure 3-28 Virtual condition rule – Problem 3.

3. When inspecting the eight-hole pattern:
   A. Does the center hole, datum B, apply at MMC or virtual condition? _____
      If the center hole were produced at Ø1.260, how much shift tolerance
      would be available from the center hole? _____
      _____

   B. Does the keyseat, datum C, apply at MMC or virtual condition? _____
      If the keyseat were produced at 0.505, how much shift tolerance would
      be available from the keyseat? _____

Datums

Datums are the reference surfaces or the starting points for the location and orientation of features. They are essential for appropriate and complete tolerancing of a part. Datum geometries can become very complicated when they are features of size, compound datums, or features of an unusual shape. Geometric dimensioning and tolerancing provides the framework necessary for dealing with these complex datums. The simpler plus or minus tolerancing system ignores these complexities, which means that plus or minus toleranced drawings cannot adequately tolerance size features. As a result, many plus or minus toleranced drawings are subject to more than one interpretation.

Chapter Objectives

After completing this chapter, you will be able to

- *Define* a datum
- *Explain* how a part is immobilized
- *Demonstrate* how datum features apply
- *Select* datum features
- *Demonstrate* the proper application of datum feature symbols
- *Demonstrate* how to specify an inclined datum feature
- *Explain* how datum planes are established on a cylindrical part
- *Explain* how datums are established
- *Explain* the application of multiple datum features
- *Demonstrate* the application of partial datum features
- *Explain* the use of datum targets

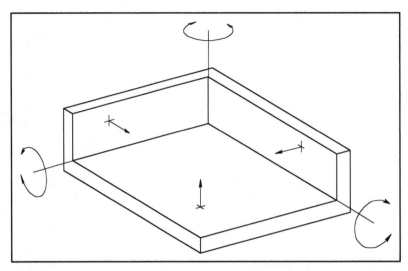

Figure 4-1   The three mutually perpendicular intersecting planes of a datum reference frame.

Definition

Datums are theoretically perfect points, lines, and planes. They establish the origin from which the location or geometric characteristics of features of a part are established. These points, lines, and planes exist within a structure of three mutually perpendicular intersecting planes known as a datum reference frame as shown in Fig 4-1.

Immobilization of a Part

Parts are thought to have six degrees of freedom, three degrees of translational freedom, and three degrees of rotational freedom. A part can move back and forth in the X direction, in and out in the Y direction, and up and down in the Z direction. It can also rotate around the X-axis, the Y-axis, and the Z-axis.

A part is oriented and immobilized relative to the three mutually perpendicular planes of the datum reference frame (as shown in Fig. 4-2) in a selected order of precedence. The datum reference frame is not absolutely perfect, but it is made sufficiently accurate with respect to the part to consider it to be perfect. Parts are relatively imperfect. In order to properly place an imperfect, rectangular part in a datum reference frame, the primary datum feature sits flat on one of the planes of the datum reference frame with a minimum of three points of contact that are not in a straight line. The secondary datum feature is pushed up against a second plane of the datum reference frame with a minimum of two points of contact. Finally, the part is slid along the first two planes of the datum reference frame until the third datum feature contacts the third plane of the datum reference frame with a minimum of one point of contact. The primary datum plane on the part contacting the datum reference frame eliminates

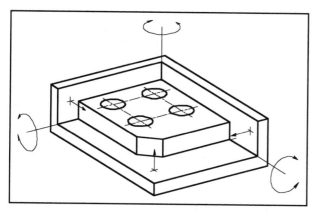

Figure 4-2 Immobilizing of a part within the three mutually
perpendicular intersecting planes of a datum reference frame.

three degrees of freedom, translation in the Z direction and rotation around the
X-axis and the Y-axis. The secondary datum plane on the part contacting the
datum reference frame eliminates two degrees of freedom, translation in
the Y direction and rotation around the Z-axis. The tertiary datum plane on
the part contacting the datum reference frame eliminates one degree of freedom, translation in the X direction.

Datums are specified in order of precedence as they appear from left to right
in the feature control frame; they need not be in alphabetical order. Datum A
in Fig. 4-3 is the primary datum, datum B is the secondary datum, and datum
C is the tertiary datum because this is the order in which they appear in the
feature control frame.

Figure 4-3 The order of precedence of datums is
determined by the order in which they appear
from left to right in the feature control frame.

Application of Datums

Measurements cannot be made from theoretical surfaces. Therefore, datums
are assumed to exist in and be simulated by the processing equipment such as
surface plates, gages, machine tables, and vises. Processing equipment is not
perfect but is made accurately enough to simulate datums. The three mutually perpendicular planes of a datum reference frame provide the origin and
direction for measurements from datums to features.

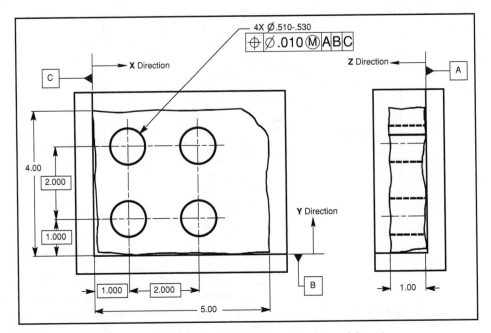

Figure 4-4   A datum reference frame provides measurement origin and direction.

Figure 4-4 shows an imperfect part placed in a relatively perfect datum reference frame. The back of the part is identified as datum A, which is specified in the feature control frame as the primary datum. In this example, the primary datum feature must make contact with the primary datum reference plane with a minimum of three points of contact; as a result, the primary datum controls the orientation—in this case perpendicularity—of features toleranced to that datum reference frame—A, B, and C. Datums B and C are the lower and left edges of the part and identified as the secondary and tertiary datums, respectively. Dimensions are measured from, and are perpendicular to, the perfect datum reference frame, not the imperfect datum features of the part.

The selection of secondary and tertiary datums depends on the characteristics of these features, such as feature size, and whether or not they are mating surfaces. However, if the two features are of the same size, do not mate with other features, and are essentially equal in every respect, then either one of them could be the secondary datum. Even though selecting a secondary datum over a tertiary datum may be arbitrary, one datum must precede the other, keeping in mind that all applicable datums must be specified. The specification of datums in order of precedence allows the part to be placed in the datum reference frame the same way every time.

Variations of form that fall within the size tolerance may occur on the datum feature. If variations on datum features fall within the size tolerance but exceed design requirements, they can be controlled with a form tolerance.

Datum Feature Selection

Datum features are selected to meet design requirements. When selecting datum features, the designer should consider the following characteristics:

- Functional surfaces
- Mating surfaces
- Readily accessible surfaces
- Surfaces of sufficient size to allow repeatable measurements

Datum features must be easily identifiable on the part. If parts are symmetrical, or have identical features making identification of datum features impossible, the datums features must be physically identified.

Selecting datums is the first step in dimensioning a part. Figure 4-4 shows a part with four holes. The designer selected the back of the part as the primary datum, datum A, because the back of the part mates with another part, and the parts are bolted together with four bolts. Datum A makes a good primary datum for the four holes because the primary datum controls orientation, and it is desirable to have bolt holes perpendicular to mating surfaces. The hole locations are dimensioned from the bottom and left edges of the part. Datum B is specified as the secondary datum, and datum C is specified as the tertiary datum in the feature control frame. Datum surfaces for location are selected because of their relative importance to the controlled features. The bottom edge of the part was selected as the secondary datum because it is larger than the left edge. The left edge might have been selected as the secondary datum if it were a mating surface.

Datum Feature Identification

All datum features must be specified. Datums may be designated with any letter of the alphabet except I, O, and Q. A datum feature symbol is used to identify physical features of a part as datum features. Datum feature symbols must not be applied to centerlines, center planes, or axes.

The datum feature symbols attached to the center planes in Fig. 4-5 are ambiguous. It is not clear whether the outside edges, one of the hole patterns, or the slots are the features that determine these center planes. The other datum feature symbols in Fig. 4-5 are attached to actual features and are acceptable

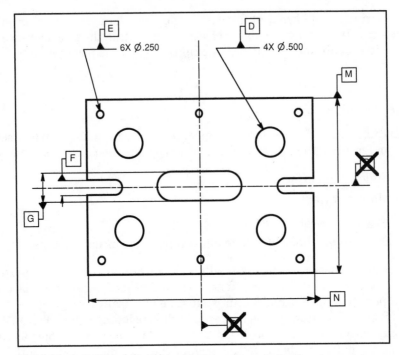

Figure 4-5   Datum feature symbols should not be applied to imaginary planes or lines.

as datums. The center planes can then be determined from actual features on the part.

Inclined Datum Features

If a surface is at an angle other than 90° to the datum reference frame, especially if the corner is rounded or broken off, it may be difficult to locate features to that surface. One method, shown in Fig. 4-6, is to place a datum feature symbol on the inclined surface and control that surface with an angularity tolerance and a basic angle. Datum features are not required to be perpendicular to each other. Only the datum reference frame is defined as three mutually perpendicular intersecting planes. To inspect this part, a precision 30° wedge is placed in a datum reference frame. The part is then placed in the datum reference frame with datum C making at least one point of contact with the 30° wedge.

Cylindrical Datum Features

Cylindrical parts might have an inside or outside diameter as a datum. A cylindrical datum feature is always associated with two theoretical planes meeting

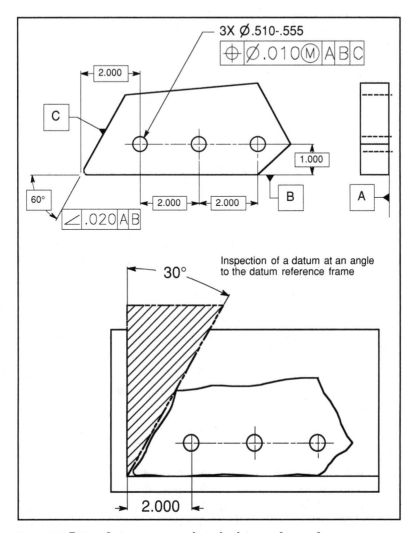

Figure 4-6 Datum features at an angle to the datum reference frame.

at right angles at its datum axis. The part in Fig. 4-7 may be mounted in a centering device, such as a chuck or a V-block, so that the center planes intersecting the datum axis can be determined. Another datum feature, in this example, datum C, may be established to control rotational orientation or clocking of the hole pattern about the datum axis.

Establishing Datums

Two kinds of features may be specified as datums:

- Features not subject to size variations such as plane flat surfaces

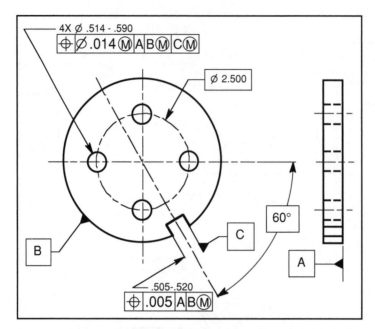

Figure 4-7 A pattern located to a cylindrical datum feature and clocked to a third datum.

- Features subject to size variations (also known as features of size and size features)

Plane flat surfaces

When features not subject to size variation, such as datums A, B, and C in Fig. 4-8, are specified as datum features, the corresponding datums are simulated by plane surfaces. Plane, flat-surface features on a rectangular-shaped part make the most convenient datums. Unfortunately, many parts are not rectangular, and designers are often forced to select datums that are features subject to size variations, such as cylinders.

Features subject to size variations at RFS

Features subject to size variations, such as datum D, are specified with one of the material condition modifiers, regardless of feature size (RFS) or maximum material condition (MMC). If a datum feature of size is specified at RFS, then processing equipment, such as gages, chucks, and mandrels, must make physical contact with the datum feature. This means that when gaging the four-hole pattern to datum hole D specified at RFS, as in feature control frame 1 in Fig. 4-8, the inspector must used the largest pin that fits through datum hole D in order to make physical contact with the hole.

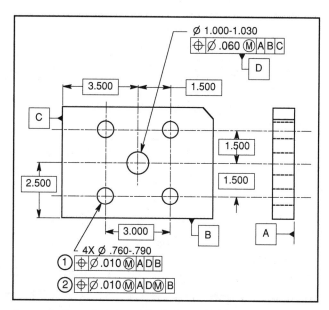

Figure 4-8 Datum planes A, B, and C and a feature of size, datum D.

Features subject to size variations at MMC

If a datum feature of size is specified at MMC, as in feature control frame 2 in Fig. 4-8, the size of the mating feature on the processing equipment has a constant boundary. The constant boundary pin is specified at the MMC of the datum feature or at its virtual condition if the virtual condition rule applies. This means that when gaging the hole pattern to a datum feature of size specified at MMC, the pin that fits through the datum hole is produced at the MMC or the virtual condition of the datum hole. Because datum hole D specified by feature control frame 2 has a geometric tolerance and is specified as a secondary datum, the virtual condition applies. See the virtual condition rule in chapter 3. The virtual condition for datum hole D is Ø .940 in. Therefore, datum D pin used to gage the four-hole pattern is Ø .940. As the datum feature departs from Ø .940 toward Ø 1.030, a shift tolerance exists about datum D in the amount of such departure. See Chapter 7 for a complete discussion of shift tolerance.

Plane flat surfaces vs. features subject to size variations

In Fig. 4-9A, the primary datum, datum A, controls the orientation of the part and must maintain a minimum of three points of contact with the top surface of the mating gage. The pin, datum B, easily assembles in the mating hole with a possible shift tolerance since it is specified at MMC. In Fig. 4-9B, the primary datum, datum A, must also maintain a minimum of three points of contact with the top surface of the mating gage, but datum B, specified at RFS,

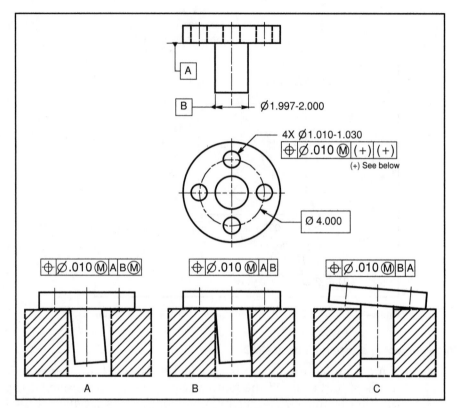

Figure 4-9   A datum feature of size specified at MMC, at RFS, and as primary and secondary datums.

must make physical contact with the gage. Therefore, the size of the hole in the gage must be adjustable to contact the surface of the pin, datum B, even if it contacts the pin only at two points. In Fig. 4-9C, the primary datum is datum B, and it is specified at RFS. Because datum B is primary, it controls the orientation of the part. Because the pin is specified at RFS, it must make physical contact and align with the hole in the gage. In this case, datum B on the gage must be adjustable not only to contact the surface of the datum B pin, but the adjustable gage must align the pin to the gage with a minimum of three points of contact. Datum A contacts the top surface of the gage at only one point.

If a datum feature symbol is in line with a dimension line, as datums B and C in Fig. 4-10, the datum is the size feature measured by that dimension. The 7.00-inch size feature between the left and right edges is datum B, and the 5.00-inch size feature between the top and bottom edges is datum C. The four-hole pattern and the Ø 3.00-inch hole are controlled to datums B and C as specified in the feature control frames. It is understood that the four-hole pattern is located to the center planes of datums B and C, and no dimensions are required from

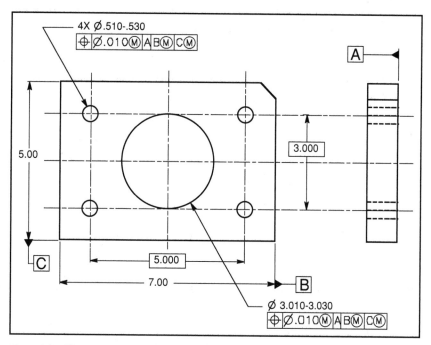

Figure 4-10 Features controlled to datum features of size.

the center planes to the pattern. The Ø 3.00-inch hole is also located on the axis intersected by the center planes of datums B and C. Since datums B and C are specified at MMC (circle M), a shift tolerance is available in each direction as each feature of size departs from MMC toward least material condition.

Multiple Datum Features

When more than one datum feature is used to establish a single datum, the datum reference letters and appropriate modifiers are separated by a dash and specified in one compartment of the feature control frame, as shown in Fig. 4-11. Together, the two datum features constitute one composite datum feature axis where datum A is no more important than datum B and datum B is no more important than datum A. When a cylinder is specified as a datum, such as datums A and B, the entire surface of the feature is considered to be the datum feature. Theoretically, the entire surface of a cylindrical datum feature is to contact the smallest precision sleeve that will fit over the cylinder. Similarly, the entire surface of an internal cylindrical datum feature is to contact the largest precision pin that will fit inside the cylinder. This almost never happens since inspectors typically do not have this kind of equipment. An external cylindrical feature is usually placed in a three-jaw chuck or in a set of V-blocks. An internal cylindrical feature is often placed on an adjustable mandrel. A part such as the

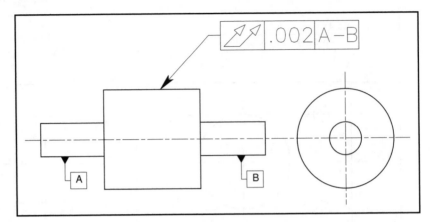

Figure 4-11   Multiple datum features A and B are of equal value.

one in Fig. 4-11 would probably be placed in a set of V-blocks to inspect the total runout specified.

A Partial Surface as a Datum Feature

When a surface is specified as a datum, the entire feature is considered to be the datum feature. If only a part of a feature is required to be the datum feature, such as datums A and B in Fig. 4-12, then a heavy chain line is drawn adjacent to the surface profile and dimensioned with basic dimensions.

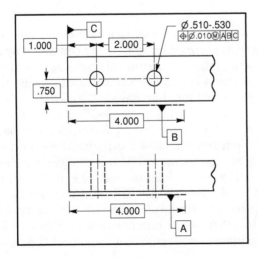

Figure 4-12   Partial datum features.

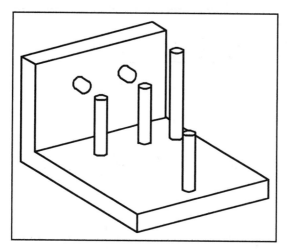

Figure 4-13 Fixture for the part shown in Fig. 4-14.

Datum Targets

Some manufacturing processes, such as casting, forging, welding, and heat treating, are likely to produce uneven or irregular surfaces. Datum targets may be used to immobilize parts with such uneven or irregular surfaces. Datum targets may also be used to support irregular-shaped parts that are not easily mounted in a datum reference frame. Datum targets are used only when necessary because, once they are specified, costly manufacturing and inspection tooling is required to process them.

Datum targets are designed to contact parts at specific points, lines, and areas. These datum targets are usually referenced from three mutually perpendicular planes to establish a datum reference frame. A primary datum plane is established by a minimum of three datum targets not in a straight line. Two datum targets are used to establish a secondary datum plane. And one datum target establishes a tertiary datum plane. A combination of datum target points, lines, and areas may be used. Datum target points are represented on a drawing by target point symbols and identified by datum target symbols such as datum targets B1 and B2 shown in Fig. 4-14. The datum target points, lines, and areas are connected to the datum target symbols with a radial line.

Actual tooling points on the fixture are not points at all but pins with hemispherical ends contacting the part with the highest point on the hemisphere, as shown in Fig. 4-14. A datum target line is represented by a target point symbol on the edge of the part in the top view of the drawing and by a phantom line on the front view, datum target C1. Since datum target C1 is on the far side of the part, a dashed radial line is used to connect the datum target symbol. Where a datum target area is required, the desired area is outlined by a phantom line and filled with section lines, as shown for datum targets A1,

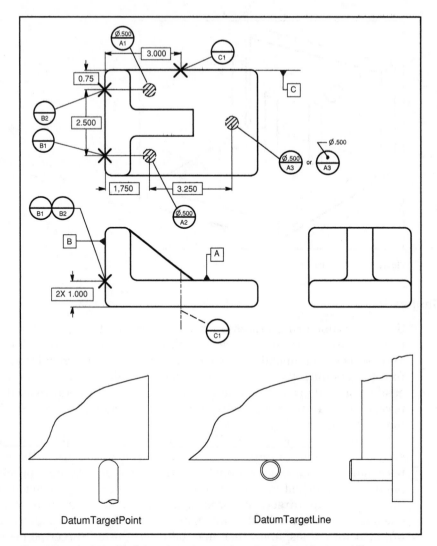

Figure 4-14   A part with datum target areas, target points, and a target line.

A2, and A3. All datum targets are dimensioned for location and size either by toleranced dimensions or basic dimensions. Basic dimensions are toleranced with gage-makers' tolerances.

Datum targets established on a cylindrical part

The axis of a primary datum feature specified at RFS may be established by two sets of three equally spaced datum targets, as shown in Fig. 4-15.

A datum target line is represented by a target point symbol on the edge view of the cylinder and a phantom line drawn across the cylinder—see datum target

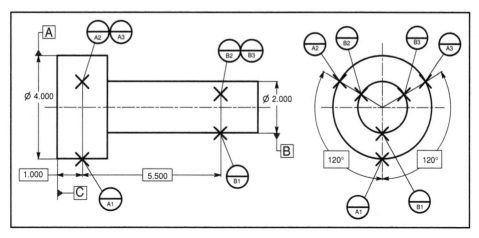

Figure 4-15 Datum targets on a cylindrical part.

line A1 in Fig. 4-16. Where a datum target area is required, the desired area is bounded by phantom lines and filled with section lines, as shown for datum target area B1.

Step and equalizing datums

A datum plane may have a step or offset such as datum A in Fig. 4-17. The step between datum points A1-A2 and A3 is specified with a basic dimension of 1.000 inch.

Equalizing datums are used to center parts that have circular ends like the part in Fig. 4-17. On this part, a 90° V-shaped knife edge, B1 and B2, and two datum target points, C1 and C2, are used to center the cylindrical ends to the fixture. Equalizing datums may be used to center other similar geometries.

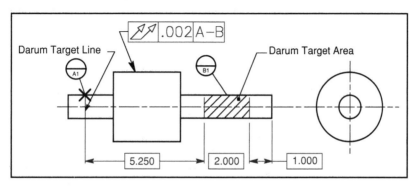

Figure 4-16 Datum target line and area on cylindrical features.

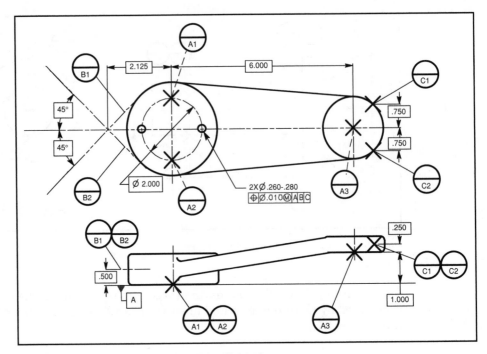

Figure 4-17 Datum targets for step and equalizing datums.

Summary

- Datums are theoretically perfect points, lines, and planes.
- Datums exist within a structure of three mutually perpendicular intersecting planes known as a datum reference frame.
- A part is oriented and immobilized relative to the three mutually perpendicular planes of the datum reference frame in a selected order of precedence.
- Since measurements cannot be made from theoretical surfaces, datums are assumed to exist in and be simulated by the processing equipment.
- Datums are specified in order of precedence as they appear in the feature control frame.
- Datum features are selected to meet design requirements. Functional surfaces, mating surfaces, readily accessible surfaces, and surfaces of sufficient size to allow repeatable measurements make good datum features.
- A datum feature symbol is used to identify physical features of a part as datum features. Datum feature symbols should not be applied to centerlines, center planes, or axes.
- A cylindrical datum feature is always intersected by two theoretical planes meeting at right angles at its datum axis.

- Plane, flat-surface features not subject to size variations make the best datums.

- When datum size features are specified at RFS, the processing equipment must make physical contact with the datum features.

- When datum features of size are specified at MMC, the size of the mating feature on the processing equipment has a constant boundary.

- When a cylinder is specified as a datum, the entire surface of the feature is considered to be the datum feature.

- Datum targets may be used to immobilize parts with uneven or irregular surfaces.

Chapter Review

1. Datums are theoretically perfect _____.

2. Datums establish the _____ from which the location or geometric characteristic of features of a part are established.

3. Datums exist within a structure of three mutually perpendicular intersecting planes known as a _____.

4. To properly position a part with datum features that are plane surfaces in a datum reference frame, the datum features must be specified in order of _____.

5. The primary datum feature contacts the datum reference frame with a minimum of _____ points of contact that are not in a straight line.

6. Datums are assumed to exist in and be simulated by the _____.

7. Datums are specified in order of precedence as they appear in the _____ _____.

8. Datums need not be in _____ order.

9. When selecting datum features, a designer should consider features that are:

   _____

   _____

   _____

   _____

10. The primary datum controls _____.

11. A datum feature symbol is used to identify _____ of a part as datum features.

12. Datum feature symbols should not be applied to _____ _____.

13. One method of tolerancing datum features at an angle to the datum reference frame is to place a datum feature symbol on the _____ and control that surface with an angularity tolerance and a basic angle.

14. A _____ is always intersected by two theoretical planes meeting at right angles at its datum axis.

15. The two kinds of features specified as datums are:

_____

_____

16. Size features may apply at _____

_____.

17. When size features are specified at RFS, the processing equipment must make _____ with the datum features.

18. When size features are specified at MMC, the size of the processing equipment has a _____.

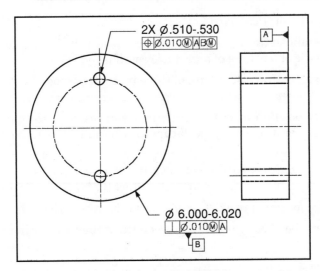

Figure 4-18 Datum feature of size drawing for questions 19–24.

19. The two-hole pattern is perpendicular to what datum?_____

20. The two-hole pattern is located to what datum? _____

21. If inspected with a gage, what is the datum B diameter of the gage?

_____

22. If inspected with a gage, what is the diameter of the two pins on the gage _____?

23. If datum B had been specified at RFS, explain how the gage would be different.

_____

_____

24. If datum B had been specified as the primary datum at RFS, explain how the gage would be different.

_____

_____

_____

25. If a datum feature symbol is in line with a dimension line, the datum is the _____ measured by the dimension.

26. When cylinders are specified as datums at RFS, the entire surface is considered to be the _____

27. When more than one datum feature is used to establish a single datum, the _____ and appropriate _____ are separated by a dash and specified in one compartment of the feature control frame.

28. If only a part of a feature is required to be the datum feature, then a _____ is drawn adjacent to the surface profile and dimensioned with basic dimensions.

29. Datum targets may be used to immobilize parts with _____.

30. Costly manufacturing and inspection _____ is required to process datum targets.

Problems

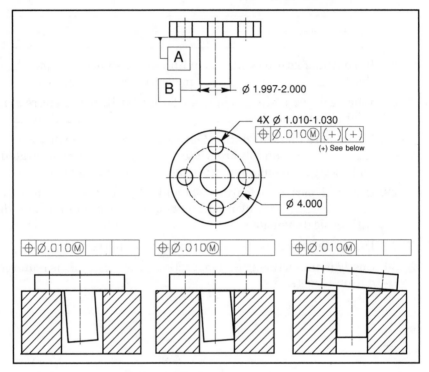

Figure 4-19   Datums at MMC and RFS: Problem 1.

1. Complete the feature control frames with datums and material condition symbols to reflect the drawing in Figure 4-19.

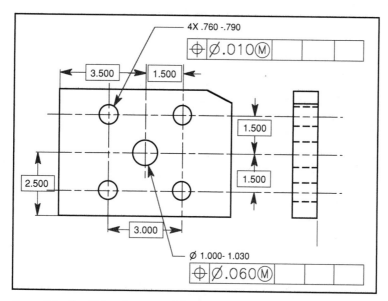

Figure 4-20 Specifying datums and datum feature symbols: Problem 2.

2. Provide the appropriate datum feature symbols on the drawing and datums in the feature control frames in the datum exercise above.

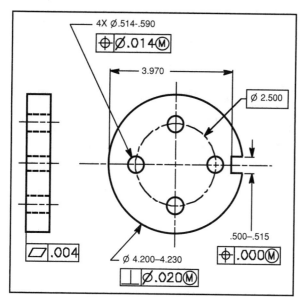

Figure 4-21 Specifying datums and datum feature symbols: Problem 3.

3. Specify the appropriate datum feature symbols and datums in the drawing in Fig. 4-21.

5

Form

All form tolerances apply to single, or individual, features; consequently, form tolerances are independent of all other features. No datums apply to form tolerances. The form of individual features is automatically controlled by the size tolerance—Rule #1. When the size tolerance does not adequately control the form of a feature, a form tolerance may be specified as a refinement. Except for straightness of a median line and of a median plane, all form tolerances are surface controls and are attached to feature surfaces with a leader or, in some cases, an extension line. No cylindrical tolerance zones or material conditions are appropriate for surface controls.

Chapter Objectives

After completing this chapter, you will be able to

- *Specify and interpret* flatness
- *Specify and interpret* straightness
- *Explain* the difference between straightness of a surface and straightness of a median line or median plane
- *Specify and interpret* circularity
- *Specify and interpret* cylindricity
- *Specify and interpret* free-state variation

Flatness

Definition

Flatness of a surface is a condition where all line elements of that surface are in one plane.

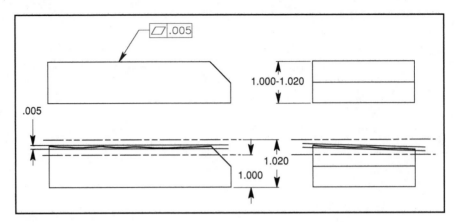

Figure 5-1 Flatness.

Specifying flatness tolerance

In a view where the surface to be controlled appears as a line, a feature control frame is attached to the surface with a leader or extension line, as shown in Fig. 5-1. The feature control frame contains a flatness symbol and a numerical tolerance. Normally, nothing else appears in the feature control frame unless unit flatness is specified, as shown below. Flatness tolerance is a refinement of the size tolerance, Rule #1, and must be less than the size tolerance. The thickness at each local size must fall within the limits of size, and the size feature may not exceed the boundary of perfect form at maximum material condition.

Interpretation.   The surface being controlled in Fig. 5-1 must lie between two parallel planes separated by the flatness tolerance of .005 specified in the feature control frame. In addition, the surface must fall within the size tolerance, the two parallel planes .020 apart. The flatness tolerance zone does not need to be parallel to any other surface as indicated in the right side view. The standard states that the flatness tolerance must be less than the size tolerance, but the size tolerance applies to both the top and bottom surfaces of the part. The manufacturer will probably use only about half of the size tolerance, producing the part in Fig. 5-1 about 1.010 thick. Since the MMC of 1.020 minus the

TABLE 5-1   Flatness Tolerances for the Part in Fig. 6-1

Actual part size	Flatness tolerance	Controlled by
1.020	.000	
1.018	.002	Rule #1
1.016	.004	
1.014	.005	
1.010	.005	Flatness Tolerance
1.005	.005	
1.000	.005	

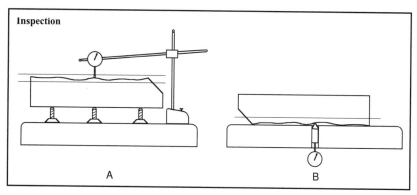

Figure 5-2 Two flatness verification techniques.

actual size of 1.010 is an automatic Rule #1 form tolerance of .010, a flatness tolerance refinement of .005, as specified in the feature control frame, seems appropriate. The entire part in Fig. 5-1 must fit between two parallel planes 1.020 apart. If the thickness of the part is produced at anywhere between 1.015 and 1.020, the form of the part, flatness, is controlled by Rule #1. If the thickness of the part if between 1.000 and 1.014, the geometric tolerance insures that the top surface of the part does not exceed a flatness of .005 as shown in Table 5.1.

Inspection.   First, the size feature is measured to verify that it falls within the limits of size. Then, flatness verification is achieved by measuring the surface, in all directions, to determine that the variation does not exceed the tolerance in the feature control frame. The measurement of surface variation in Fig. 5-2A is performed with a dial indicator after the surface in question has been adjusted with jackscrews to remove any parallelism error. In Fig. 5-2B, flatness is measured by moving the part over the probe in the surface plate and reading the indicator to determine the flatness error. This is a relatively simple method of measuring flatness; no adjustment is needed. However, specialized equipment is required.

Unit flatness

Flatness may be applied on a unit basis to prevent abrupt variations in surface flatness. The overall flatness of .010 in the feature control frame in Fig. 5-3 applies to the entire surface. The refinement of .001 per square inch applies to each and every square inch on the surface as an additional requirement to the overall flatness.

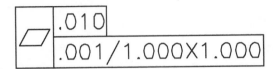

Figure 5-3 An overall flatness of .010 with unit flatness as a refinement.

Straightness

Definition

Straightness is a condition where a line element on a surface, a median line, or a line element of a median plane is a straight line.

Specifying straightness of a surface tolerance

In a view where the line elements to be controlled appear as a line, a feature control frame is attached to the surface with a leader or extension line, as shown in Fig. 5-4. The feature control frame contains a straightness symbol and a numerical tolerance. Normally, nothing else appears in a feature control frame controlling straightness of a surface unless unit straightness is specified. Straightness tolerance is a refinement of the size tolerance Rule #1 and must be less than the size tolerance. The size feature may not exceed the boundary of perfect form at MMC.

Interpretation. The line elements being controlled in Fig. 5-4 must lie between two parallel lines separated by the straightness tolerance of .004 specified in the feature control frame and parallel to the view in which they are specified—the front view. In addition, the line elements must fall within the size tolerance of .020. The straightness tolerance zone is not required to be parallel to the bottom surface or axis of the respective parts. Each line element is independent of all

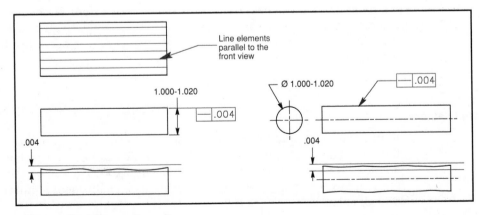

Figure 5-4 Straightness of a surface.

TABLE 5-2 Flatness Tolerances for the Part in Fig. 6-1

Actual part size	Straightness tolerance	Controlled by
1.020 MMC	.000	
1.018	.002	Rule #1
1.016	.004	
1.014	.004	
1.010	.004	Straightness Tolerance
1.005	.004	
1.000 LMC	.004	

other line elements. Straightness tolerance must be less than the size tolerance. The parts in Fig. 5-4 are likely to be produced at a thickness of 1.010 for the rectangular part and a diameter of 1.010 for the cylindrical part. Since the MMC of 1.020 minus the actual size of 1.010 is the automatic Rule #1 form tolerance of .010, a straightness tolerance refinement of .004 as specified in the feature control frame seems appropriate. The entire rectangular part in Fig. 5-4 must fit between two parallel planes 1.020 apart, and the entire cylindrical part must fit inside a cylindrical hole 1.020 in diameter. Just as for flatness, if the thickness/diameter of the parts is produced anywhere between 1.016 and 1.020, the straightness of each part is controlled by Rule #1 shown in Table 5-2.

Inspection. First, the size feature is measured to verify that it falls within the limits of size. Then, straightness verification is achieved by measuring line elements on the surface, parallel to the view in which they are specified, to determine that straightness variation does not exceed the tolerance indicated in the feature control frame. The measurement of surface variation for straightness

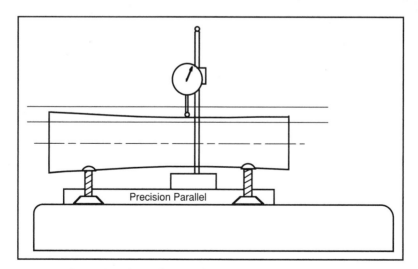

Figure 5-5 Inspection of straightness of a surface.

is performed similar to the measurement for flatness. Straightness of a cylindrical surface is inspected by moving the dial indicator across the surface plate, against the edge of a precision parallel, measuring line elements on the surface parallel to the axis of the cylinder as indicated in Fig. 5-5. Each line element is independent of every other line element, and the surface may be readjusted to remove any parallelism error for the measurement of each subsequent line element. There are an infinite number of line elements on any surface. The inspector must measure a sufficient number of line elements to be convinced that all line elements fall within the tolerance specified. Straightness verification of line elements on a flat surface is measured in a similar fashion. Parallelism error must be removed, and each line element is measured parallel to the surface on which the straightness control appears.

Specifying straightness of a median line and median plane

When a feature control frame with straightness tolerance is associated with a **size dimension**, the straightness tolerance applies to the median line of a cylinder, as in Fig. 5-6A, or a median plane for a noncylindrical feature, as in Fig. 5-6B. The median plane derived from the surfaces of the noncylindrical feature may bend, warp, or twist in any direction away from a perfectly flat plane but must not exceed the tolerance zone boundaries.

Interpretation.   While each actual local size of a feature must fall within the size tolerance, the features in Fig. 5-6 may exceed the boundary of perfect form at

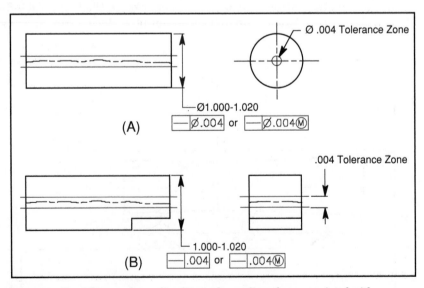

Figure 5-6 Straightness of a median line and a median plane associated with dimensions of size features.

TABLE 5-3 Straightness Tolerances for the Parts in Fig. 5-6

Feature size	Straightness tolerances			
	Cylindrical feature (Straightness of a median line)		Noncylindrical feature (Straightness of a median plane)	
	⌀ .004	⌀ .004 Ⓜ	.004	.004 Ⓜ
1.020 MMC	Ø .004	Ø .004	.004	.004
1.015	Ø .004	Ø .009	.004	.009
1.010	Ø .004	Ø .014	.004	.014
1.005	Ø .004	Ø .019	.004	.019
1.000 LMC	Ø .004	Ø .024	.004	.024

MMC because of bending or warping. A straightness control of a median line or median plane will allow the feature to violate Rule #1. Straightness associated with a size dimension may be specified at regardless of feature size (RFS) or at MMC. If specified at RFS, the tolerance applies at any increment of size within the size limits. If specified at MMC, the total straightness tolerance equals the tolerance in the feature control frame plus any bonus tolerance, equal to the amount of departure from MMC toward LMC. Consequently, a feature with a straightness control of a median line or median plane has a virtual condition. Both parts in Fig. 5-6 have a virtual condition of 1.024.

Inspection. First, a size feature is measured to verify that it falls within its limits of size. Then, straightness verification of a size feature specified at MMC can be achieved by placing the part in a full form functional gage, as shown in Fig. 5-7. If a part goes all the way in the gage and satisfies the size requirements,

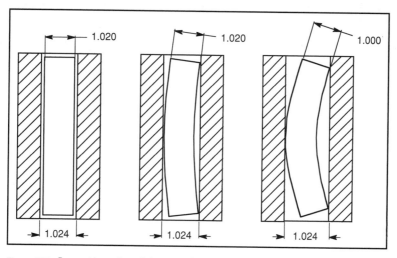

Figure 5-7 Inspection of straightness of a size feature at MMC.

it is a good part. Straightness verification of a size feature specified at RFS can be achieved by taking differential measurements on opposite sides of the part with a dial indicator to determine how much the median line varies from a perfectly straight axis or the median plane varies from a perfectly flat center plane. If the bow or warp of the part exceeds the tolerance in the feature control frame, at any size within the size tolerance, the part is not acceptable.

Circularity

Definition

Circularity (roundness) has two definitions, one for a surface of revolution about an axis and the other for a sphere. Circularity is a condition of a surface:

- For a surface of revolution, all points on the surface intersected by a plane perpendicular to the axis are equidistant from that axis.
- For a sphere, all points on the surface intersected by a plane passing through the center are equidistant from that center.

Specifying circularity tolerance

A feature control frame is attached to the surface of the feature with a leader. The leader may be attached to the surface in the circular view of a cylinder, as shown in Fig. 5-8, or it may be attached to the surface in the longitudinal view. The feature control frame contains a circularity symbol and a numerical tolerance. Normally, nothing else appears in the feature control frame. (In some cases, the *free-state* symbol is included in the feature control frame for parts subject to free-state variation.) Circularity tolerance is a refinement of the size

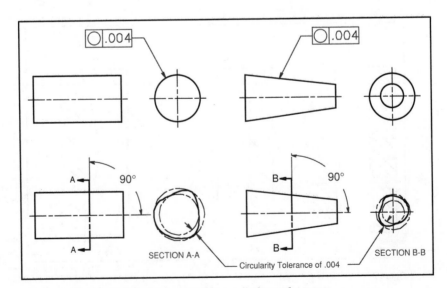

Figure 5-8   Circularity tolerance applied to a cylinder and a taper.

tolerance (Rule #1) and must be less than the size tolerance, except for parts subject to free-state variation. Rule #1 controls circularity with a diametral tolerance across the diameter, and the geometric tolerance controls circularity with a radial tolerance, so in actuality, the geometric tolerance should be less than half of the size tolerance specified on the diameter. If more information about circularity tolerance is required, a complete discussion on the subject is available in the *ANSI B89.3.1 Measurement of Out-of-Roundness*.

Interpretation.   Circular elements in a plane perpendicular to the axis of the part on the surface being controlled must lie between two concentric circles, in which the radial distance between them is equal to the tolerance specified in the feature control frame. Each circular element is independent of every other circular element. That means that the part can look like a stack of pennies that is misaligned and yet can still satisfy a circularity inspection. Rule #1 requirements limit misalignment.

Inspection.   The feature must first be measured at each cross section to determine that it satisfies the limits of size and Rule #1. Then, the part is placed on the precision turntable of the circularity inspection machine and centered with the centering screws. The probe contacts the part while it is being rotated on the turntable. The path of the probe is magnified and plotted simultaneously on the polar graph as the part rotates. The circular path plotted on the polar graph in Fig. 5-9 falls within two circular elements on the graph. This particular measurement of the part is circular within a radial distance of .002.

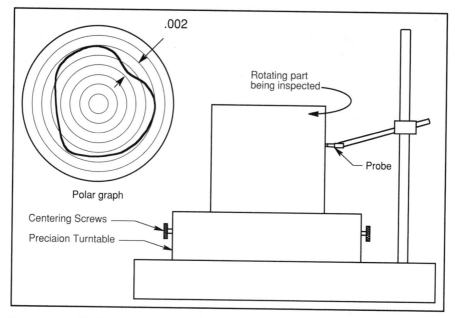

Figure 5-9 Verification of circularity with a circularity inspection machine.

Cylindricity

Definition

Cylindricity is a condition where all points on the surface of a cylinder are equidistant from the axis.

Specifying cylindricity tolerance

A feature control frame may be attached to the surface of a part with a leader in either the circular view or the rectangular view. The feature control frame contains a cylindricity symbol and a numerical tolerance. Normally, nothing else appears in the feature control frame. Cylindricity tolerance is a refinement of the size tolerance (Rule #1) and must be less than the size tolerance.

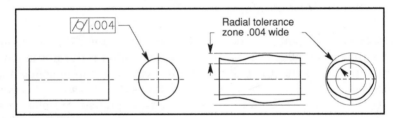

Figure 5-10   Cylindricity tolerance.

Interpretation.   The surface being controlled must lie between two coaxial cylinders in which the radial distance between them is equal to the tolerance specified in the feature control frame. Unlike circularity, the cylindricity tolerance applies to circular and longitudinal elements at the same time. Cylindricity is a composite form tolerance that simultaneously controls circularity, straightness of a surface, and taper of cylindrical features.

Inspection.   The feature is first measured at each cross section to determine that it satisfies the limits of size and Rule #1. Then, the part is placed on the precision turntable of the circularity inspection machine and centered with the centering screws. The probe contacts the part and moves vertically while the turntable is rotating. The spiral path of the probe is magnified and plotted simultaneously on the polar graph as the part rotates. The spiral path must fall within two concentric cylinders in which the radial distance between them is equal to the tolerance specified in the feature control frame.

Free-State Variation

Free-state variation is a term used to describe the distortion of a part after the removal of forces applied during the manufacturing process. This distortion is primarily due to the weight and flexibility of the part and the release of internal

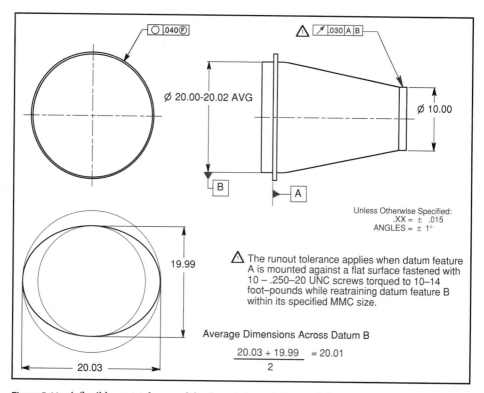

Figure 5-11 A flexible part toleranced for free-state variation and the restrained condition.

stresses resulting from fabrication. A part of this nature—for example, a large sheet metal tube or an O-ring—is referred to as a nonrigid part. A nonrigid part must meet its dimensional requirements in one of two ways, the free-state or the restrained condition.

Where a form or location tolerance is specified for a feature in the free state, the free-state symbol is placed inside the feature control frame following the tolerance and any modifiers. A size dimension and tolerance is specified followed by the abbreviation AVG indicating that the tolerance applies to the average of measurements. In Fig. 5-11, for clarity, only two measurements are shown, but a minimum of four measurements must be taken to insure the accuracy of an average diameter. If the average measurement falls inside the tolerance range, the dimension is in tolerance.

Where features are to be controlled for orientation, location, or runout in the restrained condition, the note must clearly state *which* features are to be restrained, *how* they are to be restrained, and to *what extent* they are to be restrained. Figure 5-11 contains an example of a note specifying the restrained condition for the runout control. The restrained condition should simulate actual assembly conditions.

Summary

The surface controls of flatness, straightness, circularity, and cylindricity all share the same general requirements. Straightness of a median line or median plane is quite a different control. Table 5-4 compares some of these similarities and differences.

TABLE 5-4   Summary of the Application of form Controls

	▱	—	Size feature	◯	⌭
1. Datums do not apply to these controls	X	X	X	X	X
2. Rule #1 applies to these controls	X	X		X	X
3. This is a surface control	X	X		X	X
4. This control is specified with a leader	X	X		X	X
5. This tolerance is a refinement of the size tolerance	X	X		X	X
6. This tolerance violates Rule #1			X		
7. This is a size feature control			X		
8. This control is associated with the dimension			X		
9. This form may exceed the size tolerance			X		
10. The Ø symbol and circle M symbol may be used			X		

Chapter Review

1. Form tolerances are independent of all _____.

2. No _____ apply to form tolerances.

3. The form of individual features is automatically controlled by the _____ _____.

4. A form tolerance may be specified as a refinement when _____ _____.

5. All form tolerances are surface controls except for _____ _____.

6. No _____ or _____ are appropriate for surface controls.

7. Flatness of a surface is a condition where all line elements on that surface are in one _____.

8. In a view where the surface to be controlled with a flatness tolerance appears as a _____ a feature control frame is attached to the surface with a _____.

9. The feature control frame controlling flatness contains a _____ and a _____.

10. The surface being controlled for flatness must lie between _____ separated by the flatness tolerance. In addition, the feature must fall within the _____.

11. The flatness tolerance zone does not need to be _____ to any other surface.

12. The size feature may not exceed the _____ _____.

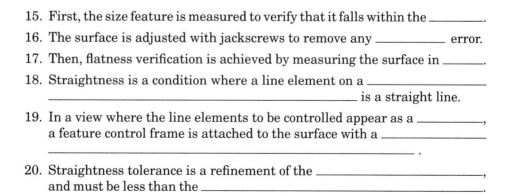

Figure 5-12   Specifying flatness.

13. Specify the flatness of the top surface of the part in Fig. 5-12 within .006 in a feature control frame.

14. Draw a feature control frame with an overall flatness of .015 and a unit flatness of .001 per square inch.

15. First, the size feature is measured to verify that it falls within the _____.

16. The surface is adjusted with jackscrews to remove any _____ error.

17. Then, flatness verification is achieved by measuring the surface in _____.

18. Straightness is a condition where a line element on a _____ _____ is a straight line.

19. In a view where the line elements to be controlled appear as a _____, a feature control frame is attached to the surface with a _____ _____ .

20. Straightness tolerance is a refinement of the _____, and must be less than the _____.

TABLE 5-5   Problem 21

Actual part size	Straightness tolerance	Controlled by
1.020		
1.018		
1.016		
1.014		
1.010		
1.005		
1.000		

21. Complete Table 5.5 above specifying the straightness tolerance and what controls it for the drawing in Fig. 5-4.

22. The measurement of surface variation for straightness is performed similar to the measurement for _____.

23. Each line element is _____ of every other line element.

24. When a feature control frame with a straightness tolerance is associated with a size dimension, the straightness tolerance applies to _____
_____ .

25. While each actual local size must fall within the size _____, the feature controlled with straightness of a median line or median plane may exceed the _____
at MMC.

26. A straightness control of a median line or median plane will allow the feature to violate _____.

27. If specified at MMC, the total straightness tolerance of a median line or median plane equals the tolerance in the feature control frame plus any
_____ .

TABLE 5-6   Problem 28

	Cylindrical feature	
	(Straightness of a median line)	
Feature size	—⊘.006	—⊘.006 Ⓜ
1.020 MMC		
1.015		
1.010		
1.005		
1.000 LMC		

28. Complete Table 5.6 above specifying the appropriate tolerances for the sizes given.

29. Straightness verification of a size feature specified at MMC can be achieved by _____ .

30. Straightness verification of a size feature specified at _____ cannot be achieved by placing the part in a full form functional gage.

31. Circularity tolerance consists of two _____ in which the _____ between them is equal to the tolerance specified in the feature control frame.

32. For circularity verification, the feature must first be measured at each cross section to determine that it satisfies the _____ and _____ .

33. Circularity can be accurately inspected on a _____ .

34. Cylindricity is a condition where all points on the surface of a cylinder are _____ .

35. The cylindricity tolerance consists of two _____ in which the _____ between them is equal to the tolerance specified in the _____ .

36. Cylindricity is a _____ form tolerance that simultaneously controls _____ of cylindrical features.

37. On Table 5-7, place an X under the control that agrees with the statement.

TABLE 5-7   Problem 37

	▱	—	Size feature	O	⌀
1. Datums do not apply to these controls					
2. This tolerance violate Rule #1					
3. This is a size feature control					
4. This control is associated with the dimension					
5. This tolerance may exceed the size tolerance					
6. Rule #1 applies to this tolerance					
7. This tolerance is a surface control					
8. This control is specified with a leader					
9. This tolerance is a refinement of Rule #1					
10. The Ø, circle M, and circle L symbols may be used					

38. *Free-state variation* is a term used to describe the distortion of a part after the removal of forces applied during the _____ .

39. Where a form or location tolerance is specified for a feature in the free state, the free-state symbol is placed inside the _____ following the _____ . A minimum of _____ must be taken to insure the accuracy of an average diameter.

40. A minimum of _____ must be taken to insure the accuracy of an average diameter.

41. The restrained condition should simulate _____ .

Problems

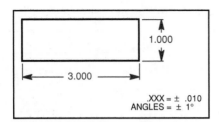

Figure 5-13   Flatness: Problem 1.

1. Specify a flatness control of .005 for the top surface of the part in Fig. 5-13.

2. Draw a feature control frame with a unit flatness of .003 per square inch and an overall flatness of .015.

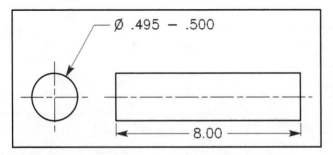

Figure 5-14   Straightness of a surface: Problem 3.

3. Specify straightness of a surface of .002 on the cylinder in the drawing in Fig. 5-14.

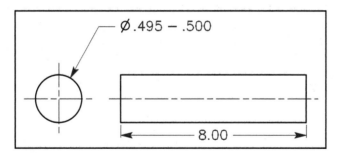

Figure 5-15 Straightness of a median line—Problem 4.

4. Specify straightness of a median line of .010 at MMC on the cylinder in the drawing in Fig. 5-15.

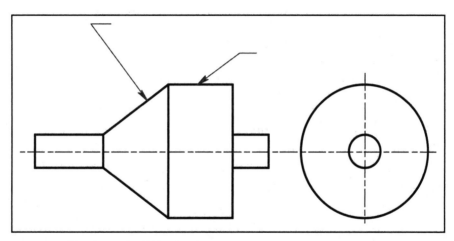

Figure 5-16 Circularity: Problems 5 and 6.

5. Specify a circularity tolerance of .002 on the cone in the drawing in Fig. 5-16.

6. Specify a cylindricity tolerance of .0005 on the cylinder in the drawing in Fig. 5-16.

6

Orientation

Orientation is the general term used to describe the angular relationship between features. Orientation controls include parallelism, perpendicularity, angularity, and, in some cases, profile. All orientation controls must have datums. It makes no sense to specify a pin, for instance, to be perpendicular. The pin must be perpendicular to some other feature. The other feature is the datum.

Chapter Objectives

After completing this chapter, you will be able to

- *Specify* tolerances that will control flat surfaces parallel, perpendicular, and at some basic angle to datum features
- *Specify* tolerances that will control axes parallel, perpendicular, and at some basic angle to datum features

The orientation of a plane surface controlled by two parallel planes and an axis controlled by a cylindrical tolerance zone will be discussed in this chapter. When a plane surface is controlled with a tolerance zone of two parallel planes, the entire surface must fall between the two planes. Since parallelism, perpendicularity, angularity, and profile control the orientation of a plane surface with a tolerance zone of two parallel planes, they also control flatness if a flatness tolerance is not specified. When it is desirable to control only the orientation of individual line elements of a surface, a note, such as EACH ELEMENT or EACH RADIAL ELEMENT, is placed beneath the feature control frame.

When an axis is controlled by a cylindrical tolerance zone, the entire axis must fall inside the tolerance zone. Although axes and center planes of size features may be oriented using two parallel planes, in most cases, they will be controlled by other controls, such as a position control, and will not be discussed in this chapter. The position control is a composite control, which controls location and orientation at the same time. Parallelism, perpendicularity, and angularity are often used to refine the orientation of other controls such as the position control.

Parallelism

Definition

Parallelism is the condition of a surface or center plane, equidistant at all points from a datum plane; also, parallelism is the condition of an axis, equidistant along its length from one or more datum planes or a datum axis.

Specifying parallelism of a flat surface

In a view where the surface to be controlled appears as a line, a feature control frame is attached to the surface with a leader or extension line, as shown in Fig. 6-1. The feature control frame contains a parallelism symbol, a numerical tolerance, and at least one datum. The datum surface is identified with a datum feature symbol. Parallelism tolerance of a flat surface is a refinement of the size tolerance, Rule #1, and must be less than the size tolerance. The size feature may not exceed the maximum material condition (MMC) boundary, and the thickness at each actual local size must fall within the limits of size.

Interpretation. The surface being controlled in Fig. 6-1 must lie between two parallel planes separated by the parallelism tolerance of .005 specified in the feature control frame. The tolerance zone must also be parallel to the datum plane. In addition, the surface must fall within the size tolerance, the two parallel planes .020 apart. The entire part in Fig. 6-1 must fit between two parallel planes 1.020 apart. The controlled surface may not exceed the boundary of

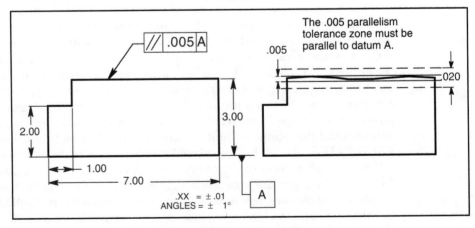

Figure 6-1 Specifying a plane surface parallel to a plane surface.

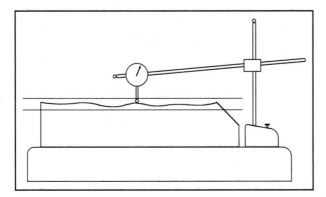

Figure 6-2 Verifying parallelism of a flat surface.

perfect form at MMC, Rule #1. Parallelism is the only orientation control that, where applied to a flat surface, requires a perfect angle (parallelism is a 0° angle) at MMC. Since the parallelism control applies to a surface, no material condition symbol applies.

Inspection.   Verifying the parallelism of a flat surface is relatively easy. First, the size feature is measured to determine that it falls within the limits of size. Next, the datum surface is placed on top of the surface plate. Then, verification is achieved, as shown in Fig. 6-2, by using a dial indicator to measure the surface in all directions to determine that any variation does not exceed the tolerance specified in the feature control frame.

Specifying parallelism of an axis

When controlling the parallelism of a size feature, the feature control frame is associated with the size dimension of the feature being controlled. In Fig. 6-3, the feature control frame is attached to the extension of the dimension line. The feature control frame contains a parallelism symbol, numerical tolerance, and at least one datum. If the size feature is a cylinder, the numerical tolerance is usually preceded by a diameter symbol, as shown in Fig. 6-3. There are some cases where an axis is controlled by two parallel planes, but these are very uncommon and would probably be toleranced with the position control. The tolerance and the datum in the feature control frame both apply to size features, and they apply regardless of feature size (RFS) since no material condition symbol is specified. The datum feature is identified with a datum feature symbol.

If the tolerance and the datum both apply at MMC, as in Fig. 6-4, then the tolerance has a possible bonus tolerance, and the datum has a possible shift tolerance. Bonus and shift tolerances will both be discussed in more detail in the chapter on position.

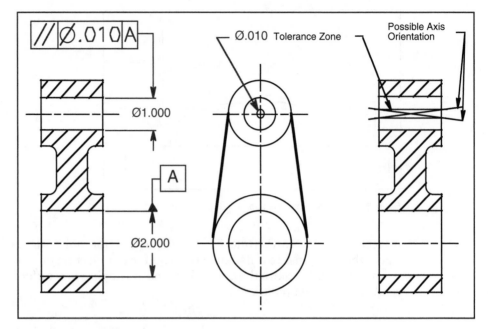

Figure 6-3   Controlling one axis parallel to another axis.

Figure 6-4   The parallelism tolerance and datum both applied at MMC.

Perpendicularity

Definition

Perpendicularity is the condition of a surface, axis, or center plane that is at a 90° angle to a datum plane or datum axis.

Specifying perpendicularity of a flat surface

In a view where the surface to be controlled appears as a line, a feature control frame is attached to the surface with a leader or extension line, as shown in Fig. 6-5. The feature control frame contains a perpendicularity symbol, a numerical tolerance, and at least one datum. The datum feature is identified with a datum feature symbol.

Interpretation. The surface being controlled must lie between two parallel planes separated by the perpendicularity tolerance of .010 specified in the feature control frame. Also, the tolerance zone must be perpendicular to the datum

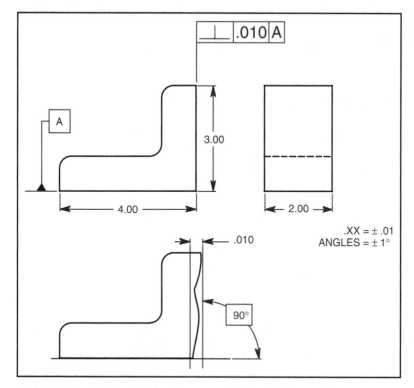

Figure 6-5 Specifying a plane surface perpendicular to a datum plane.

plane. All size features of the part must fall within the limits of size and may not exceed the boundary of perfect form at MMC, Rule #1. There is no boundary of perfect *orientation* at MMC for perpendicularity. The 90° angles on the part also have a tolerance. The title block angularity tolerance controls all angles, including 90° angles, which are not otherwise toleranced. Since the perpendicularity control applies to a surface, no material condition symbol applies.

Inspection.   The datum surface is clamped on an angle plate that sits on a surface plate. Then, as shown in Fig. 6-6, perpendicularity verification is achieved by using a dial indicator to measure the surface in all directions to determine that any variation does not exceed the tolerance specified in the feature control frame.

Tangent plane

The *tangent plane* symbol (circle T) in the feature control frame specifies that the perpendicularity tolerance applies to the precision plane contacting the high points of the surface. Even though the surface irregularities exceed the perpendicularity tolerance, if a precision plane contacting the high points of a surface falls inside the specified tolerance zone, the surface is in tolerance.

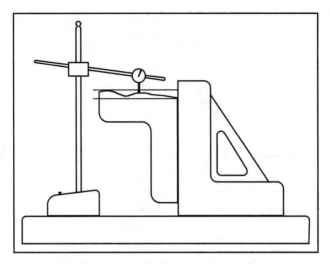

Figure 6-6   Verifying perpendicularity of a flat surface.

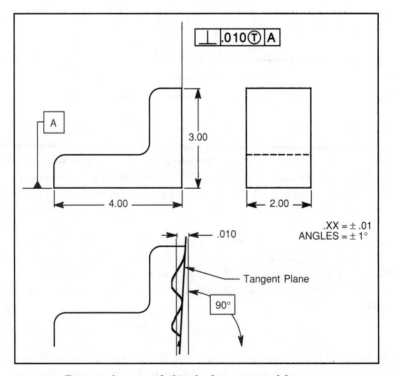

Figure 6-7   Tangent plane specified in the feature control frame.

The tangent plane of the toleranced surface in Fig. 6-7 lies inside the tolerance zone. The tangent plane concept allows the acceptance of more parts.

Specifying perpendicularity of an axis

When controlling the perpendicularity of a size feature, the feature control frame is associated with the size dimension of the feature being controlled. The feature control frame contains a perpendicularity symbol, a numerical tolerance, and at least one datum. If the size feature is a cylinder, the numerical tolerance is usually preceded by a diameter symbol, as shown in Fig. 6-8. A cylindrical tolerance zone that controls an axis perpendicular to a plane surface, such as the drawing in Fig. 6-8, is perpendicular to that surface in all directions around the axis. There are some cases where an axis is controlled by two parallel planes, but these are very uncommon and would probably be toleranced with the position control. The perpendicularity tolerance may be larger or smaller than the size tolerance. Since the tolerance in the feature control frame applies to the pin, a size feature, and no material condition symbol is specified, RFS applies. If the tolerance applies at MMC, as in Fig. 6-9, then a possible bonus tolerance exists. The datum feature is identified with a datum feature symbol.

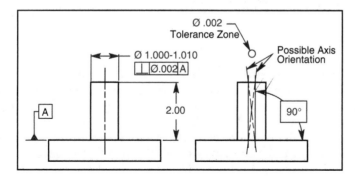

Figure 6-8   Specifying an axis perpendicular to a datum plane.

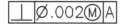

Figure 6-9   The perpendicularity tolerance applied at MMC.

Angularity

Definition

Angularity is the condition of a surface, axis, or center plane at a specified angle other than parallel or perpendicular to a datum plane or datum axis.

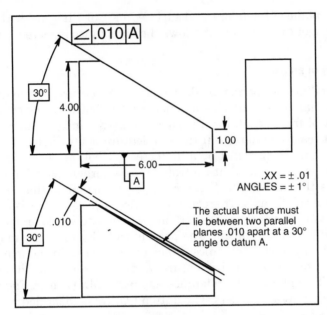

Figure 6-10 Specifying an angularity tolerance for a plane surface at a basic angle to a plane surface.

Specifying angularity of a flat surface

In a view where the surface to be controlled appears as a line, a feature control frame is attached to the surface with a leader or extension line. If an extension line is used, it needs to only contact the feature control frame at a corner, as shown in Fig. 6-10. The feature control frame contains an angularity symbol, a numerical tolerance, and at least one datum. The numerical tolerance for the surface being controlled is specified as a linear dimension because it generates a uniform-shaped tolerance zone. A plus or minus angularity tolerance is not used because it generates a nonuniform, fan-shaped tolerance zone. The datum feature is identified with a datum feature symbol.

Interpretation. The surface being controlled in Fig. 6-10 must lie between two parallel planes separated by the angularity tolerance of .010 specified in the feature control frame. The tolerance zone must be at the specified basic angle of 30° to the datum plane. All size features of the part must fall within the limits of size and may not exceed the boundary of perfect form at MMC, Rule #1. There is no boundary of perfect orientation at MMC for angularity. The 90° angles on the part also have a tolerance. The title block angularity tolerance controls all angles, including 90° angles, unless otherwise specified.

Since the angularity control applies to a surface, no material condition symbol applies.

Inspection.   The datum surface may be placed on a sine plate. The sine plate sits on a surface plate at an accurate 30° angle produced by a stack of gage blocks. The basic angle between the tolerance zone and datum A is assumed to be perfect. Inspection equipment is not perfect, but inspection instrument error is very small compared to the geometric tolerance. As shown in Fig. 6-11, once the datum surface is positioned at the specified angle, angularity verification is achieved by using a dial indicator to measure the surface in all directions to determine that any variation does not exceed the tolerance specified in the feature control frame.

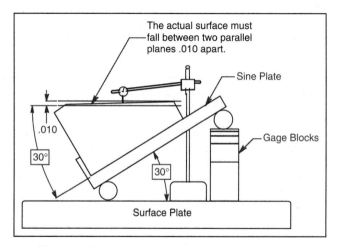

Figure 6-11   Verification of a surface at a 30° angle to a flat datum surface.

Specifying angularity of an axis

When controlling the angularity of a size feature, the feature control frame is associated with the size dimension of the feature being controlled. The feature control frame contains an angularity symbol, a numerical tolerance, and at least one datum. If the size feature is a cylinder, the numerical tolerance may or may not be preceded by a diameter symbol, as shown in Fig. 6-12. If the diameter symbol precedes the numerical tolerance, the axis is controlled with a cylindrical tolerance zone. If there is no diameter symbol preceding the numerical tolerance, the axis is controlled by two parallel planes. The tolerance in the feature control frame applies to the hole—a size feature—and it applies

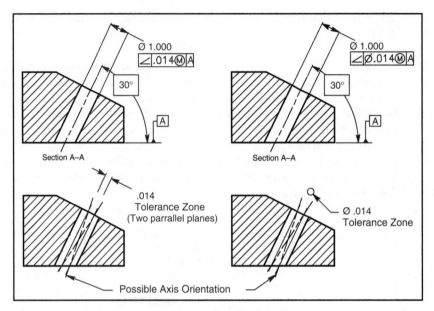

Figure 6-12 Specifying an axis at an angle to a datum plane.

at RFS since no material condition symbol is specified. The datum feature is identified with a datum feature symbol.

Figure 6-13 The angularity tolerance specified at MMC.

If the tolerance applies at MMC, as in Fig. 6-13, it has a possible bonus tolerance. When MMC or the least material condition (LMC) is desirable, it might be more appropriate to specify angularity and location at the same time by using a position control. If the design requires the angularity tolerance to be smaller than the location tolerance, the angularity tolerance at MMC can be specified as a refinement of the position tolerance at MMC, as shown in Fig. 6-14.

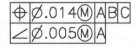

Figure 6-14 The angularity tolerance specified at MMC as a refinement to the position control.

Summary

TABLE 6-1 Orientation Summary

	Plane surfaces			Axes and control planes		
	//	⊥	∠	//	⊥	∠
Datums required	X	X	X	X	X	X
Controls flatness if flatness is not specified	X	X	X			
Circle T modifier can apply	X	X	X			
Tolerance specified with a leader or extension line	X	X	X			
May not exceed boundary of perfect form at MMC	X					
Tolerance associated with a dimension				X	X	X
Material condition modifiers apply				X	X	X
A virtual condition applies				X	X	X

Chapter Review

1. Orientation is the general term used to describe the _____ relationship between features.

2. Orientation controls include _____.

3. All orientation controls must have _____.

4. In a view where the surface to be controlled appears as a line, a feature control frame is attached to the surface with a _____.

5. The feature control frame for parallelism of a surface must contain at least _____.

6. The datum feature is identified with a _____.

7. Parallelism tolerance of a flat surface is a refinement of the size tolerance and must be less than the _____.

8. Size features may not exceed the _____.

9. A surface being controlled with a parallelism tolerance must lie between _____ separated by the parallelism tolerance specified in the feature control frame. The tolerance zone must also be _____ to the datum plane.

10. The controlled surface may not exceed the _____.

11. Parallelism is the only orientation control that, where applied to a flat surface, requires a perfect angle (parallelism is a 0° angle) at _____.

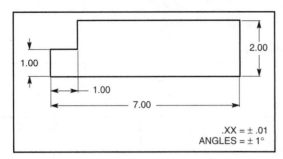

Figure 6-15 Specifying parallelism.

12. Supply the appropriate geometric tolerance on the drawing in Fig. 6-15 to control the top surface of the part parallel to the bottom surface within .010.

13. When controlling the parallelism of a size feature, the feature control frame is associated with the _____ of the feature being controlled.

14. If the size feature is a cylinder, the numerical tolerance is usually preceded by a _____.

15. A surface being controlled with a perpendicularity tolerance must lie between _____ separated by the perpendicularity tolerance specified in the feature control frame. The tolerance zone must also be _____ to the datum plane.

16. A tangent plane symbol (circle T) in the feature control frame specifies that the tolerance applies to the precision plane contacting the _____of the surface.

17. When controlling the perpendicularity of a size feature, the feature control frame is associated with the _____ of the feature being controlled.

18. If the tolerance in the feature control frame applies to a size feature and no material condition symbol is specified, _____ applies.

19. If the tolerance applies at MMC, then a possible _____ tolerance exists.

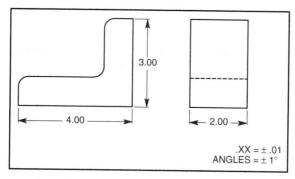

Figure 6-16 Specifying perpendicularity of a surface.

20. Supply the appropriate geometric tolerance on the drawing in Fig. 6-16 to control the 3.000-inch vertical surface of the part perpendicular to the bottom surface within .005.

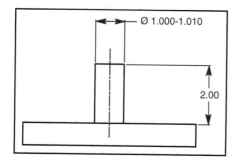

Figure 6-17 Specifying perpendicularity of a size feature.

21. Supply the appropriate geometric tolerance on the drawing in Fig. 6-17 to control the Ø 1.000-inch vertical pin perpendicular to the bottom surface of the plate within .005 at RFS.

Figure 6-18 Perpendicularity specified at MMC.

22. If the pin in Fig. 6-17 were produced at a diameter of 1.004 and toleranced with the feature control frame in Fig. 6-18, what would the total perpendicularity tolerance be? _____

23. The numerical tolerance for angularity of a surface is specified as a linear dimension because it generates a _____ zone.

24. A plus or minus angularity tolerance is not used because it generates a _____-shaped tolerance zone.

25. When controlling the angularity of a size feature, the feature control frame is associated with the _____ of the feature being controlled.

26. If the diameter symbol precedes the numerical tolerance, the axis is controlled with a _____ zone.

27. When MMC or LMC is desirable, it might be more appropriate to specify angularity and location at the same time with the _____.

TABLE 6-2 Orientation Problem

	Plane surfaces			Axes and control planes		
	//	⊥	∠	//	⊥	∠
Datums required						
Controls flatness if flatness is not specified						
Circle T modifier can apply						
Tolerance specified with a leader or extension line						
May not exceed boundary of perfect form at MMC						
Tolerance associated with a dimension						
Material condition modifiers apply						
A virtual condition applies						

28. In Table 6-2, mark an **X** in the box to indicate which control applies to the statements on the left.

Problems

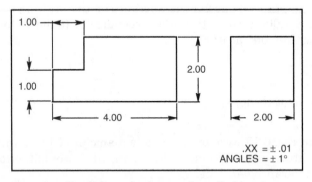

Figure 6-19 Parallelism of a plane surface—Problem 1.

1. In Fig. 6-19, specify the top surface of the part parallel to the bottom surface within a tolerance of .004. Draw and dimension the tolerance zone.

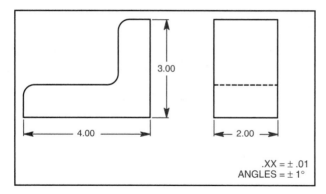

Figure 6-20 Perpendicularity of a plane surface—Problem 2.

2. In Fig. 6-20, specify the 3.000-inch surface of the part perpendicular to the bottom and back surfaces within a tolerance of .010. Draw and dimension the tolerance zone.

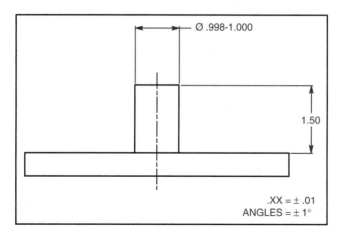

Figure 6-21 Perpendicularity of a pin to a plane surface—Problem 3.

3. In Fig. 6-21, specify the 1.000-inch pin perpendicular to the top surface of the plate within a tolerance of .015 at MMC. On the drawing, sketch and dimension a gage used to inspect this part.

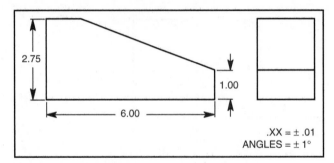

Figure 6-22   Angularity of a plane surface—Problem 4.

4. In Fig. 6-22, specify the top surface of the part to be at an angle of 20° to the bottom surface within a tolerance of .003. Draw and dimension the tolerance zone.

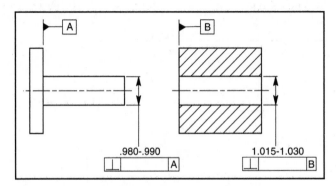

Figure 6-23   Orientation—Problem 5.

5. In Fig. 6-23, complete the feature control frames so that the two parts will always assemble, datums A and B will meet, and the part can be produced using the most cost-effective design. The pin is machined in a lathe, and the hole is drilled.

Position, General

Position is a composite tolerance that controls both the location and the orientation of size features at the same time. It is the most frequently used of the 14 geometric characteristics. The position tolerance significantly contributes to part function, part interchangeability, optimization of tolerance, and communication of design intent.

Chapter Objectives

After completing this chapter, you will be able to

- *Specify* position tolerance for the location of a size feature
- *Interpret* tolerance specified at the regardless of feature size (RFS) condition
- *Calculate* bonus and shift tolerances for features specified at the maximum material condition (MMC)
- *Specify* position tolerance and calculate the minimum wall thickness at the least material condition (LMC)
- *Calculate* boundary conditions
- *Calculate* tolerances specified with zero positional tolerance at MMC

Definition

The tolerance of position may be viewed in either of the following two ways:

- A *theoretical tolerance zone* located at true position of the toleranced feature within which the center point, axis, or center plane of the feature may vary from true position

- A *virtual condition boundary* of the toleranced feature, when specified at MMC or LMC and located at true position, which may not be violated by its surface or surfaces

Specifying the Position Tolerance

Since the position tolerance controls only size features, such as pins, holes, tabs, and slots, the feature control frame is always associated with a size dimension. In Fig. 7-1, the hole is located and oriented with the position control. In this case, the feature control frame is placed under the local note describing the diameter and size tolerance of the hole. The location of true position of this hole, the theoretically perfect location of the axis, is specified with basic dimensions from the datums indicated in the feature control frame. Once the feature control frame is assigned, an imaginary tolerance zone is defined and located about true position. The datum surfaces have datum feature symbols identifying them. Datums A, B, and C identify the datum reference frame in which the part is to be positioned for processing.

Interpretation The feature control frame is a sentence in the GD&T language; it must be specified correctly in order to communicate design intent. The feature control frame in Fig. 7-1 tells the location tolerancing story for the hole in this part: it has a cylindrical tolerance zone .010 in diameter, the full length of the feature, specified at RFS, is perfectly perpendicular to datum plane A, located a basic 2.000 inches up from datum B, and a basic 3.000 inches over from datum

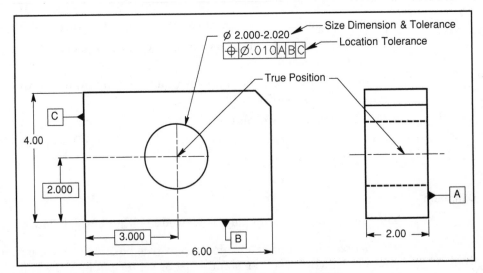

Figure 7-1 Location of a size feature with a position tolerance at RFS.

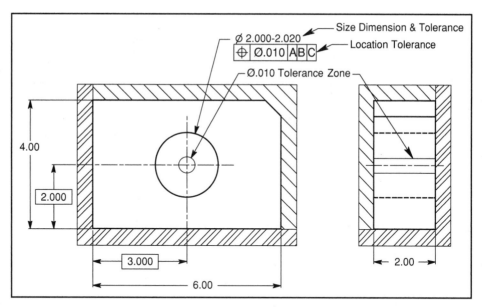

Figure 7-2 The part is placed in a datum reference frame.

C. Tolerance zones are theoretical and do not appear on drawings. A tolerance zone has been shown here for illustration purposes only.

Inspection.   Inspection starts with measuring the hole diameter. If the diameter measures 2.012, it is within the size tolerance, Ø 2.000–2.020. The next step is to measure the hole location and orientation. The part is clamped in a datum reference frame by bringing a minimum of three points on the surface of the primary datum feature into contact with the primary datum plane, a minimum of two points on the surface of the secondary datum feature into contact with the secondary datum plane, and a minimum of one point on the surface of the tertiary datum feature into contact with the third datum plane. Next, the largest pin gage to fit inside the hole is used to simulate the actual mating envelope. The actual mating envelope for an internal feature of size is the largest, similar, perfect feature counterpart that can be inscribed within the feature so that it just contacts the surface of the hole at the highest points. As shown in Fig. 7-3, the distance from the surface plate, datum B, to the top of the pin gage is measured. Measurements are also taken along the pin gage to determine that the hole is within the perpendicularity tolerance to the angle plate, datum A. Suppose the distance from the surface plate to the top of the pin is 3.008. That measurement minus half of the diameter of the pin gage equals the distance from datum B to the actual axis of the hole, $3.008 - (2.012/2) =$ 2.002. The distance, then, from true position to the actual axis of the hole in the vertical direction is .002. With the part still clamped to it, the angle plate is rotated 90°, and the distance from datum C to the actual axis of the hole is

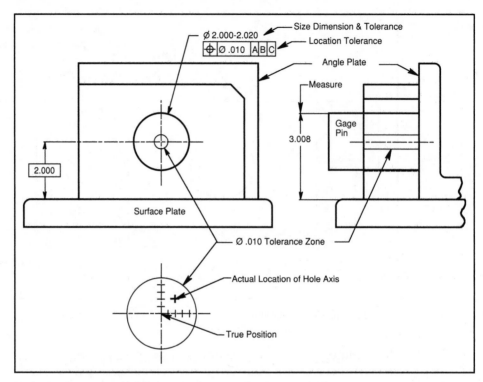

Figure 7-3   Inspecting the hole location by using the theoretical tolerance zone.

measured by repeating the previous measurement procedure. If the distance from true position to the actual axis in the horizontal direction is .002, the actual axis is .002 up and .002 over from true position requiring a tolerance zone diameter of less than .006 in diameter, well within the .010 diameter cylindrical tolerance zone shown in Fig. 7-3. The hole is within tolerance.

Regardless of Feature Size

RFS automatically applies for features of size where no material condition symbol is specified. Since no material condition symbol is specified in the feature control frame in Fig. 7-1, the RFS modifier automatically applies to the location and orientation of the hole. In other words, the position tolerance is Ø.010 no matter what size the hole happens to be. The feature size may be anywhere between a diameter of 2.000 and 2.020, and the tolerance remains Ø .010. No bonus tolerance is allowed.

Where datum features of size are specified at RFS, the datum is established by physical contact between the surface(s) of the processing equipment and the surface(s) of the datum feature. There is no shift tolerance for datum features specified at RFS. A holding device that can be adjusted to fit the size of the

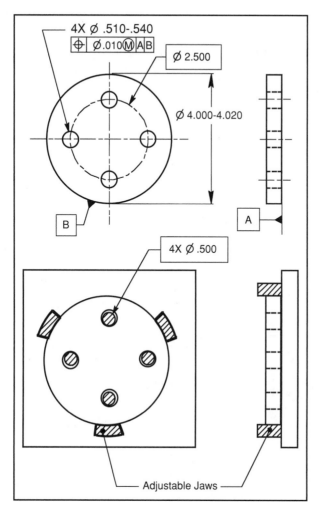

Figure 7-4 Inspecting the hole pattern controlled to a datum feature of size at RFS.

datum feature, such as a chuck, vise, or adjustable mandrel, is used to position the part. In Fig. 7-4, the outside diameter, datum B, is specified at RFS. The pattern of features is inspected by placing the outside diameter in a chucking device and the hole pattern over a set of virtual condition pins. If the part can be set inside this gage and all the feature sizes are within size tolerance, the pattern is acceptable.

Maximum Material Condition

The only difference between the tolerances in Fig.7-3 and Fig.7-5 is the MMC modifier specified after the numerical tolerance in the feature control frame.

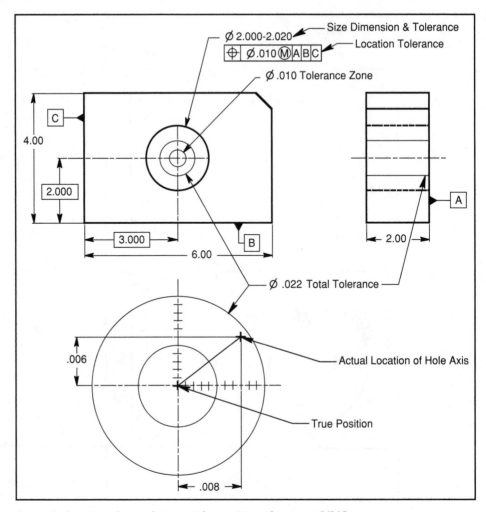

Figure 7-5   Location of a size feature with a position tolerance at MMC.

When the MMC symbol, circle M, is specified to modify the tolerance of a size feature in a feature control frame, the following two requirements apply:

1. The specified tolerance applies at the MMC of the feature. The MMC of a size feature is the largest shaft and the smallest hole. The MMC modifier, circle M, is not to be confused with the MMC of a size feature.

2. As the size of the feature departs from MMC toward LMC, a bonus tolerance is achieved in the exact amount of such departure. Bonus tolerance equals the difference between the actual feature size and the MMC of the feature. The bonus tolerance is added to the geometric tolerance specified in the

feature control frame. MMC is the most common material condition used and is often used when parts are to be assembled.

Suppose the Ø 2.000-hole in Fig. 7-5 is inspected; the actual diameter is found to be 2.012, and the actual axis is found to be .006 up and .008 over from true position. By applying the Pythagorean theorem to these coordinates, it is easily determined that the actual axis is .010 away from true position. To be acceptable, this part requires a cylindrical tolerance zone centered on true position of at least .020 in diameter. The tolerance is only Ø .010, but there is an MMC modifier; consequently, bonus tolerance is available. The following formulas are used to calculate the bonus tolerance and total positional tolerance at MMC:

- Bonus equals the difference between the actual feature size and MMC.

- Bonus plus geometric tolerance equals total positional tolerance.

TABLE 7-1   The Calculation of Bonus Tolerance

Actual feature size	− MMC	= Bonus	Geometric + tolerance	Total positional = tolerance
2.012	2.000	.012	.010	.022

When the calculations in Table 7-1 are completed, the total positional tolerance is .022—sufficient tolerance to make the hole in the part in Fig. 7-5 acceptable.

Another way of inspecting the hole specified at MMC is with a functional gage shown in Fig. 7-6. A functional gage for this part is a datum reference frame with a virtual condition pin positioned perpendicular to datum A, located a basic 2.000 inches up from datum B and a basic 3.000 inches over from datum C. If the part can be set over the pin and placed against the datum reference frame in the proper order of precedence, then the hole is in tolerance.

A functional gage represents the worst-case mating part. It is very convenient when a large number of parts are checked or when inexperienced operators are required to check parts. Dimensions on gage drawings are either toleranced or basic. The tolerance for basic dimensions is the gage-makers' tolerance. The gage-makers' tolerance is usually only about 10 percent of the tolerance on the part. All of the tolerance for the gage comes out of the tolerance for the part. In other words, a gage may not accept a bad part, but it can reject a marginally good part.

Shift Tolerance

Shift tolerance is allocated to a feature or a pattern of features, as a group, and equals the amount a datum feature of size departs from MMC or virtual

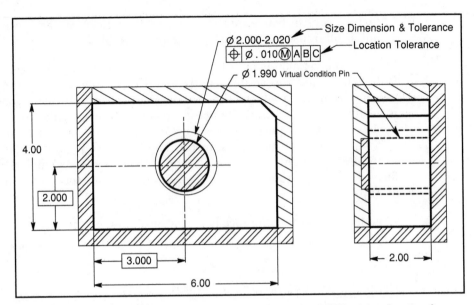

Figure 7-6 Inspecting a size feature with a position tolerance at MMC using a functional gage.

condition toward LMC. It should be emphasized that when a shift tolerance applies to a pattern of features, it applies to the pattern as a group.

Shift tolerance should not be confused with bonus tolerance. Bonus tolerance is the difference between the actual feature size and its MMC. Bonus tolerance for a particular feature is added directly to the geometric tolerance to equal the total tolerance for that feature.

Shift tolerance for a single feature of size, i.e., one feature not a pattern of features, located or oriented to a datum feature of size may be added directly to the location or orientation tolerance just like the bonus tolerance. However, shift tolerance for a pattern of features may not be added to the geometric tolerance of each feature. Treating shift tolerance for a pattern of features like bonus tolerance is a common error and should not be done for patterns of features.

Where a datum feature of size is specified with an MMC symbol, such as datum B in the feature control frames controlling the four-hole patterns in Fig. 7-7, the datum feature of size either applies at MMC or at virtual condition. As the actual size of datum feature B departs from MMC toward LMC, a shift tolerance, of the pattern as a group, is allowed in the exact amount of such departure. The possible shift equals the difference between the actual size of the datum feature and the inside diameter of the gage, as you can see in the drawings in Fig. 7-7.

In Fig. 7-7A, datum B satisfies the requirements for the virtual condition rule. In view of the fact that the perpendicularity tolerance is an orientation control, it is used to calculate the virtual condition. The virtual condition rule states that where a datum feature of size is controlled by a geometric tolerance and is

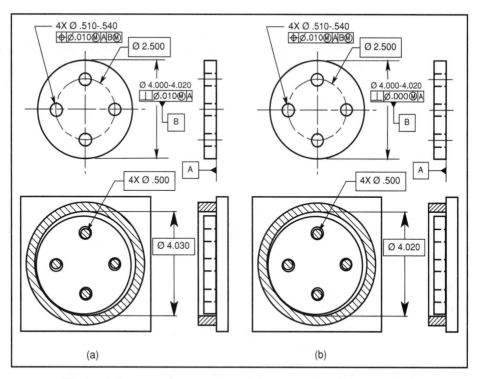

Figure 7-7 The four-hole pattern, as a group, can shift an amount equal to the difference between the sizes of the outside diameter of the part and the inside diameter of the gage.

specified as a secondary or tertiary datum, the datum applies at its virtual condition with respect to orientation. In Fig. 7-7A, the outside diameter of the part

- Is a datum, datum B
- Is a size feature
- Has a geometric tolerance, which controls orientation
- Is specified as a secondary datum in the feature control frame controlling the four-hole pattern

Virtual condition calculations

	A	B
MMC	4.020	4.020
Plus geometric tolerance (Perpendicularity)	+.010	+.000
Virtual condition (Orientation)	4.030	4.020

Because datum B on the part applies at Ø 4.030, datum B on the gage is produced at Ø 4.030. If datum feature B, on a part, is actually produced at a diameter of 4.010, the four-hole pattern, as a group, can shift .020 in any direction inside the 4.030 diameter gage, as shown in Fig.7-7A. If other inspection techniques are used, the axis of datum B, and consequently the four-hole pattern, can shift within a cylindrical tolerance zone Ø.020 centered on true position of datum B. (See the chapter on Graphic Analysis for the inspection procedure of a pattern of features controlled to a feature of size.)

In Fig. 7-7B, datum B also satisfies the requirements for the virtual condition rule, but because the perpendicularity control has a .000 tolerance, the virtual condition is the same as the MMC, Ø 4.020; consequently, datum B on the gage in Fig. 7-7B is produced Ø 4.020. If datum feature B, on a part, is actually produced at a diameter of 4.010, it can shift only .010 in any direction inside the 4.020 diameter gage, as shown in Fig.7-7B.

If a datum feature of size violates the virtual condition rule, the datum on the gage is produced at MMC. Not using geometric controls is one way to violate the virtual condition rule, but the lack of geometric controls makes it difficult to know how to make the gage.

Least Material Condition

When the LMC symbol, circle L, is specified to modify the tolerance of a size feature, the following two requirements apply:

1. The specified tolerance applies at the LMC of the feature. The LMC of a size feature is the smallest shaft and the largest hole. The LMC modifier, circle L, is not to be confused with the LMC of a size feature.

2. As the size of the feature departs from LMC toward MMC, a bonus tolerance is achieved in the exact amount of such departure. Bonus tolerance equals the difference between the actual feature size and the LMC of the feature. The bonus tolerance is added to the geometric tolerance specified in the feature control frame. LMC is the least used of the three material condition modifiers. It is often used to maintain a minimum wall thickness or maintain a minimum distance between features. The LMC modifier is just opposite in its effects to the MMC modifier. Even the form requirement of a size feature at LMC is opposite the form requirement at MMC. When a tolerance is specified with an LMC modifier, the feature may not exceed the boundary of perfect form at LMC. Finally, features toleranced at LMC cannot be inspected with functional gages. Virtual condition for internal features at LMC is equal to LMC plus the geometric tolerance.

The calculation for the virtual condition of the holes in Fig. 7-8:

LMC   1.390
Plus geometric tolerance +.010
Virtual condition @ LMC 1.400

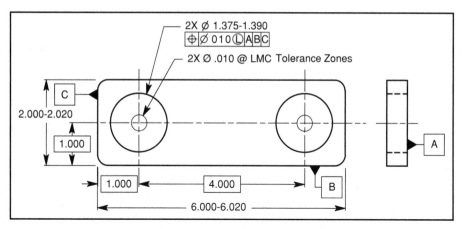

Figure 7-8 Size features toleranced with the LMC modifier.

It is not possible to put a 1.400 virtual condition pin through a 1.390 hole. Inspection of features specified at LMC must be done in some way other than with functional gages.

Calculation of Wall Thickness

What is the minimum distance between the holes and the ends of the part in Fig.7-8?

The distance from datum C to the first hole axis	1.000
Half the diameter of the hole @ LMC	− .695
Half the tolerance of the hole @ LMC	− .005
The minimum wall thickness	.300

The length of the part @ LMC	6.000
The distance from datum C to the second hole axis	− 5.000
Half the diameter of the hole @ LMC	− .695
Half the tolerance of the hole @ LMC	− .005
The minimum wall thickness	. 300

Boundary Conditions

To satisfy design requirements, it is often necessary to determine the maximum and minimum distances between features. The worst-case inner and outer boundaries, or loci, are the virtual conditions and the extreme resultant conditions; they are beneficial in performing a tolerance analysis.

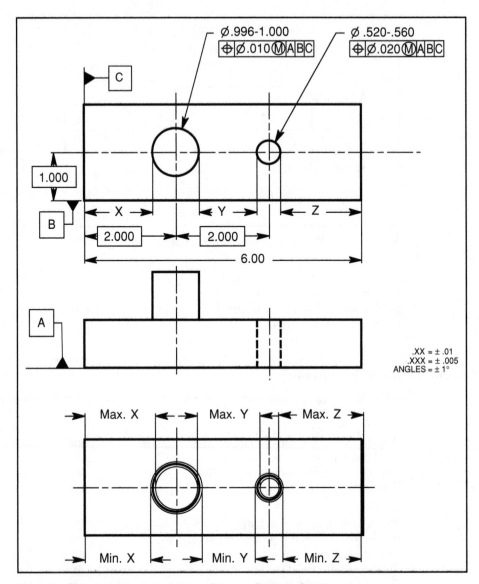

Figure 7-9 The maximum and minimum distances between features.

Calculate the maximum and minimum distances for the dimensions X, Y, and Z in Fig. 7-9. Start by calculating the virtual and the resultant conditions.

The **Virtual Condition** of the **PIN**:
$\text{V.C.}_P = \text{MMC} + \text{Geo. Tol.} =$
$\text{V.C.}_P = 1.000 + .010 =$
$\text{V.C.}_P = 1.010$ **$\text{V.C.}_P/2 = .505$**

The **Virtual Condition** of the **HOLE**:
$\text{V.C.}_H = \text{MMC} - \text{Geo. Tol.} =$
$\text{V.C.}_H = .520 - .020 =$
$\text{V.C.}_H = .500$ **$\text{V.C.}_H/2 = .250$**

Resultant Condition of the PIN:

$R.C._P = LMC - Geo. Tol. - Bonus =$

$R.C._P = .996 - .010 - .004 =$

$\textbf{R.C._P = .982} \qquad \textbf{R.C._P/2 = .491}$

Resultant Condition of the HOLE:

$R.C._H = LMC + Geo. Tol. + Bonus =$

$R.C._H = .560 + .020 + .040 =$

$\textbf{R.C._H = .620} \qquad \textbf{R.C._H/2 = .310}$

The maximum and minimum distances for dimension X:

$X_{max} = Location - R.C._P/2 =$

$X_{max} = 2.000 - .491 =$

$\textbf{X}_{\textbf{max}} = \textbf{1.509}$

$X_{min} = Location - V.C._P/2 =$

$X_{min} = 2.000 - .505 =$

$\textbf{X}_{\textbf{min}} = \textbf{1.495}$

The maximum and minimum distances for dimension Y:

$Y_{max} = Location - R.C._P/2 - V.C._H/2$

$Y_{max} = 2.000 - .491 - .250 =$

$\textbf{Y}_{\textbf{max}} = \textbf{1.259}$

$Y_{min} = Location - V.C._P/2 - R.C._H/2$

$Y_{min} = 2.000 - .505 - .310 =$

$\textbf{Y}_{\textbf{min}} = \textbf{1.185}$

The maximum and minimum distances for dimension Z:

$Z_{max} = Length_{MMC} - Loc. - V.C._H/2 =$

$Z_{max} = 6.010 - 4.000 - .250 =$

$\textbf{Z}_{\textbf{max}} = \textbf{1.760}$

$Z_{min} = Length_{LMC} - Loc. - R.C._H/2 =$

$Z_{min} = 5.990 - 4.000 - .310 =$

$\textbf{Z}_{\textbf{min}} = \textbf{1.680}$

Zero Positional Tolerance at MMC

Zero positional tolerance at MMC is just what it says—no tolerance at MMC. However, there is bonus tolerance available. As the size of a feature departs from MMC toward LMC, the bonus tolerance increases; consequently, the location tolerance is directly proportional to the size of the feature as it departs from MMC toward LMC.

Which has more tolerance, the drawing in Fig. 7-10A with a typical plus or minus tolerance for clearance holes or the drawing in Fig. 7-10B with a zero positional tolerance? It is often assumed that a zero in the feature control frame means that there is no tolerance. This misconception occurs because the meaning of the MMC symbol in the feature control frame is not clearly understood.

Zero tolerance is never used without an MMC or LMC symbol. Zero at RFS would, in fact, be zero tolerance no matter at what size the feature is produced. When zero positional tolerance at MMC is specified, the bonus tolerance applies. In many cases, the bonus is larger than the tolerance that might otherwise be specified in the feature control frame. An analysis of the part in Fig. 7-10B indicates that the holes can be produced anywhere between Ø.500 and Ø.540 . If the holes are actually produced at Ø.530, the total location tolerance available is a cylindrical tolerance zone of .030. The actual hole size, .530, minus the MMC, .500, equals a bonus tolerance of .030. Geometric dimensioning and tolerancing reflects the exact tolerance available. For drawing A, the hole sizes must be between Ø .525 and Ø .535. If the holes are actually produced at

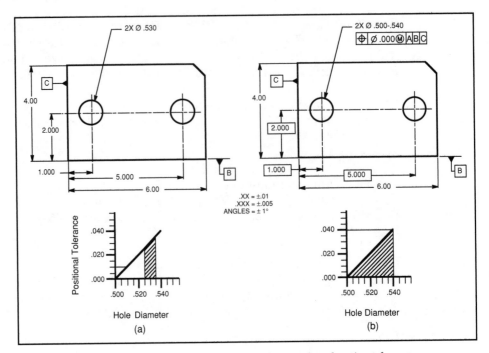

Figure 7-10   Zero positional tolerance compared to a plus or minus location tolerance.

Ø.530, the total location tolerance available is actually a cylindrical tolerance zone of .030 just as it was above. But, since the general tolerance is specified at ± .005, the inspector can accept the part only if hole locations fall within the .010 square tolerance zone specified. In this case, a tolerance of approximately .020 in each direction is wasted. Tolerance is money. How much do you want to waste?

The two parts in Fig. 7-11 are identical; they are just toleranced differently. If a part is made with the holes produced at Ø .530, what is the total location tolerance for the hole? Inspect the part by using the tolerances in drawings A and B in Fig. 7-11.

For a given hole size, the total tolerance and the virtual condition is the same whether a numerical tolerance or a zero tolerance is specified. But, the range of the hole size has been increased when zero positional tolerance is used. Some engineers do not use zero positional tolerancing at MMC because they claim that people in manufacturing will not understand it. Consequently, they put some small number such as .005 in the feature control frame with a possible .015 or .020 bonus tolerance available. If the machinists cannot read the bonus, they will produce the part within the .005 tolerance specified in the feature control frame and charge the company for the tighter tolerance. If zero positional tolerance is used, suppliers either will not bid on the part or will ask what it means. Actually, machinists who understand how to calculate bonus

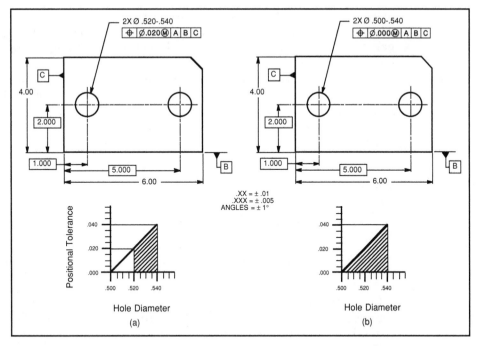

Figure 7-11   A specified position tolerance compared to zero positional tolerance.

tolerance really like the flexibility it gives them. Inspection can easily accept more parts reducing manufacturing costs.

Suppose a part is to be inspected with the drawing in Fig. 7-11A. The part has been plated a little too heavily, and the actual size of both holes is Ø .518. The inspector has to reject the part because the holes are too small. Suppose both

TABLE 7-2 Both the Total Positional Tolerance and the Virtual Condition are the Same Whether Controlled with a Nominal Tolerance or Zero Positional Tolerance

Total Positional Tolerance		
	Drawing **A**	Drawing **B**
Actual Hole Size	.530	.530
Maximum Material Condition	−.520	−.500
Bonus	.010	.030
Geometric Tolerance	+.020	+.000
Total Tolerance	**.030**	**.030**
Virtual Condition		
	Drawing **A**	Drawing **B**
Maximum Material Condition	.520	.500
Geometric Tolerance	−.020	−.000
Virtual Condition	**.500**	**.500**

holes were located within a cylindrical tolerance zone of .010, would the part assemble? The answer to that question can be determined by inspecting the part to the equivalent, zero positional toleranced drawing in Fig. 7-11B. The hole size of Ø .518 is acceptable since it falls between Ø .500 and Ø .540. The hole size, Ø .518, minus MMC, Ø .500, equals a total cylindrical tolerance of .018. The part will fit and function since only a tolerance of Ø .010 is needed. How many good parts do you want to scrap? If you find this to be a continuing problem for a particular part, you might want to submit an engineering change order converting the tolerance to a zero positional tolerance.

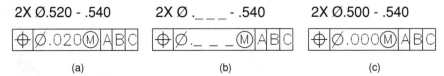

Figure 7-12   Converting the positional tolerance of a hole to a zero positional tolerance.

Converting the Ø .020 positional tolerance for the holes in Fig. 7-12A to a zero positional tolerance, Fig. 7-12C is fairly simple. The only elements to be changed are the MMC and the geometric tolerance shown as blanks in Fig. 7-12B. The tolerance in the feature control frame is always converted to zero at MMC. You must make sure there is the MMC symbol, circle M, following the tolerance. Then, convert the MMC of the feature to the virtual condition. In this case, the .520 MMC minus the .020 geometric tolerance equals the virtual condition of .500.

Zero tolerance is not used when the tolerance applies at RFS or when no bonus tolerance is available as in a tolerance specified for threads or press fit pins.

Summary

- The tolerance of position may be viewed as a theoretical tolerance zone or a virtual condition boundary.

- The location of true position is specified with basic dimensions from the datums indicated. Once the feature control frame is assigned, an imaginary tolerance zone is defined and located about true position.

- RFS automatically applies to the location and orientation tolerances of the feature being controlled if no material condition modifier is specified.

- When the MMC symbol modifies a position tolerance:

  1. The tolerance applies at MMC
  2. As the size of the feature departs from MMC toward LMC, a bonus tolerance is achieved in the exact amount of such departure

- A shift tolerance is allocated to a feature or a pattern of features, as a group, and equals the amount a datum feature of size departs from MMC or virtual condition toward LMC. It should be emphasized that when a shift tolerance applies to a pattern of features, it applies to the pattern as a group.
- When the LMC symbol modifies a position tolerance:
  1. The tolerance applies at LMC
  2. As the size of the feature departs from LMC toward MMC, a bonus tolerance is achieved in the exact amount of such departure
- The worst-case inner and outer boundaries, or loci, are the virtual and the resultant conditions; they are beneficial in performing a tolerance analysis.
- Zero positional tolerancing gives machinists more flexibility because manufacturing can easily accept more parts and charge less. For a given feature size, the total tolerance and the virtual condition are the same whether a numerical tolerance or a zero tolerance is specified.

Chapter Review

1. Position is a composite tolerance that controls both the _____ _____ of size features at the same time.

2. The tolerance of position may be viewed in either of two ways:
   A _____
   _____
   _____.
   A _____
   _____
   _____.

3. Since the position tolerance controls only size features, such as pins, holes, tabs, and slots, the feature control frame is always associated with a _____.

4. The location of true position, the theoretically perfect location of an axis, is specified with _____ from the datums indicated.

5. Once the feature control frame is assigned, an imaginary_____ _____ is defined and located about true position.

5. Datum surfaces have _____ identifying them.

6. Datums A, B, and C identify a _____; _____ consequently, they describe how the part is to be held for____.

7. To inspect a hole, the largest pin gage to fit inside the hole is used to simulate the _____ _____.

8. The measurement from the surface plate to the top of the pin gage minus half of the diameter of the pin gage equals the distance from _____ .

9. If no material condition symbol is specified in the feature control frame, the _____ modifier automatically applies to the tolerance of the feature.

10. When the MMC symbol is specified to modify the tolerance of a size feature, the following two requirements apply:
   - The specified tolerance applies at _____ .
   - As the size of the feature departs from MMC toward LMC, _____ .

11. _____ equals the difference between the actual feature size and MMC

12. Bonus plus the geometric tolerance equals _____ .

Ø.510 – .550

⊕ | Ø.010Ⓜ | A | B | C

Figure 7-13 Geometric tolerance.

13. If the tolerance in Fig. 7-13 is for a pin of Ø .530, what is the total tolerance?

14. What would be the size of the hole in a functional gage to inspect this pin?

15. If the tolerance in Fig. 7-13 is for a hole of Ø .540, what is the total tolerance?

16. What would be the size of the pin on a functional gage to inspect this hole?

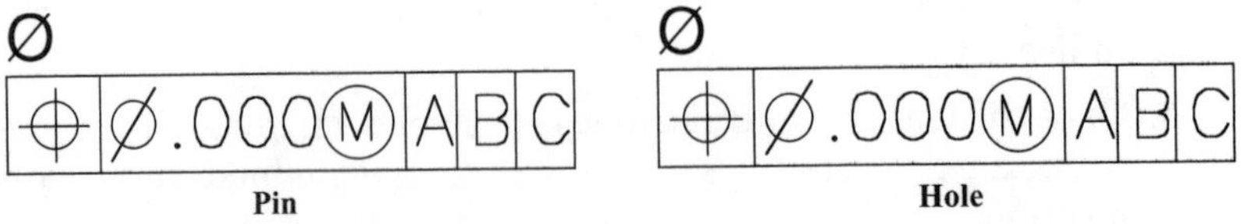

Ø   Ø

⊕ | Ø.000Ⓜ | A | B | C   ⊕ | Ø.000Ⓜ | A | B | C

Pin   **Hole**

Figure 7-14 Zero positional tolerance conversion.

17. Convert the tolerance in Fig. 7-13 to the zero positional tolerances in Fig. 7-14.

18. A shift tolerance is allocated to a feature or a pattern of features, as a group, and equals the amount a datum feature of size departs from _____ or_____.

19. When a datum feature of size is specified with the MMC symbol: the _____ applies at its MMC or virtual condition.
As the actual size of a datum feature departs from MMC toward LMC, a ___ _____, of the pattern as a group, is allowed in the exact amount of such departure.

20. The_____ states that where a datum feature of size is controlled by a geometric tolerance and is specified as a secondary or tertiary datum, the datum applies at its virtual condition with respect to orientation.

Problems

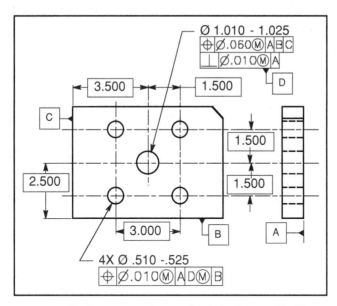

Figure 7-15 Design a gage to inspect for shift tolerance: Problem 1.

1. On a gage designed to control the four-hole pattern in Fig. 7-15, what size pin must be produced to inspect the center hole (datum D)?_____
On the same gage, what is the diameter of the four pins locating the hole pattern? _____

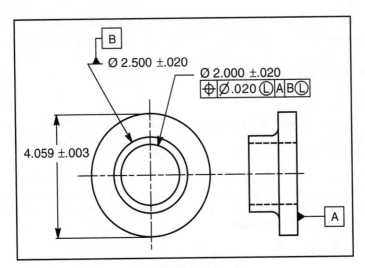

Figure 7-16   A hole specified at LMC: Problem 2.

2. Calculate the minimum wall thickness between the inside diameter and datum B in Fig. 7-16.

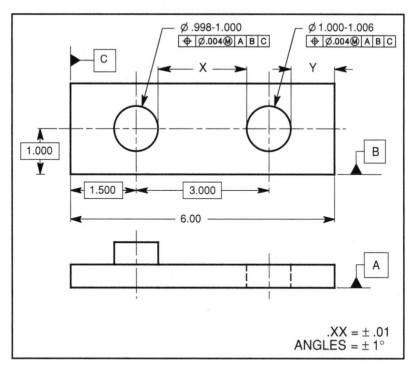

Figure 7-17   Boundary conditions: Problem 3.

3. First calculate the virtual and resultant conditions for the pin and the hole. Then calculate the maximum and minimum distances for dimensions X and Y in Fig 7-17.

The Virtual Condition of the **PIN**:   **Virtual Condition** of the **HOLE**:

Resultant Condition of the **PIN**:   **Resultant Condition** of the **HOLE**:

The maximum and minimum distances for dimension X:

$X_{\text{Max}} =$ $X_{\text{Min}} =$

The maximum and minimum distances for dimension Y:

$Y_{\text{Max}} =$ $Y_{\text{Min}} =$

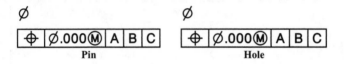

Figure 7-18 Zero positional tolerance conversion: Problem 4.

4. Convert the tolerance in Fig. 7-17 to the zero positional tolerances in Fig. 7-18. Zero tolerance is not used when the tolerance applies at _____, or when no bonus tolerance is available as in a tolerance specified for_____
_____.

Position, Location

The most important function of the position control is to locate features relative to datums and to one another. The position control is one of the most versatile of the 14 geometric controls. It controls both the location and the orientation of size features and allows the application of maximum material condition (circle M), least material condition (circle L) to features being controlled and to datum features of size. Most of the major applications of the position control are discussed in this chapter. Even though coaxiality is the location of one feature to another toleranced with the position control, it is a separate topic and will be discussed in the next chapter.

Chapter Objectives

After completing this chapter, you will be able to

- *Calculate* tolerances for floating and fixed fasteners
- *Specify* projected tolerance zones
- *Apply* the concept of multiple patterns of features
- *Demonstrate* the proper application of composite tolerancing
- *Demonstrate* the proper application of two single-segment feature control frames
- *Tolerance* nonparallel holes
- *Tolerance* counterbores
- *Tolerance* noncircular features at MMC
- *Tolerance* symmetrical features at MMC

Floating Fasteners

Because of the large number of fasteners used to hold parts together, toleranc-ing threaded and clearance holes may be one of the most frequent tolerancing activities that an engineer performs. Often, due to ignorance, habit, or both, fasteners are toleranced too tightly. This section on fasteners attempts to pro-vide the knowledge that allows engineers to make sound tolerancing decisions for floating and fixed fasteners.

The floating fastener got its name from the fact that the fastener is not re-strained by any of the members being fastened. In other words, all parts being fastened together have clearance holes in which the fastener can float before being tightened. The floating fastener formula is

$$T = H - F \qquad \text{or} \qquad H = F + T$$

Where T is the tolerance at MMC, H is the hole diameter at MMC, and F is the fastener diameter at MMC, the nominal size of the fastener. The tolerance derived from this formula applies to each hole in each part.

The floating fastener formula is simple to remember. The hole has to be larger than the fastener. The difference between the sizes of the hole and the fastener is the location tolerance, as shown graphically in Fig. 8-1.

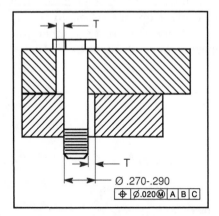

Figure 8-1 Floating fastener.

$$H = F + T$$
$$= .250 + .020$$
$$= .270$$

Once the fastener and the tolerance have been selected, it is a simple matter to calculate the MMC hole diameter. All too often, many designers simply use a reference chart for tolerancing fasteners and have little understanding of how these numbers are derived. If there is doubt about which tolerance to use, specify zero at MMC. Zero positional tolerance will provide all of the tolerance

Ø .250-.290

Figure 8-2 Floating fastener with a zero positional tolerance at MMC.

available and give the machinist the maximum size flexibility in producing the clearance hole. The calculations could not be easier. The MMC hole size when toleranced with a zero positional tolerance is the same as the diameter of the fastener.

$$H = .250 + .000 = .250$$

What is the actual location tolerance in Fig. 8-2? The location tolerance for a given hole size at MMC is the same no matter what tolerance is specified in the feature control frame. If the clearance hole is actually produced at Ø .285, the total location tolerance is:

$$\text{Geometric tolerance} + \text{bonus} = \text{total positional tolerance}$$

$$.020 + (.285 - .270) = .035$$

or

$$.000 + (.285 - .250) = .035$$

If the machinist happens to produce the hole at Ø .265 and zero positional tolerance is specified, the hole size is acceptable, but the hole must be within a location tolerance of Ø .015. No matter what tolerance is selected, it is important to use the formula to determine the correct MMC hole diameter. If the MMC clearance hole diameter is incorrect, either a possible no fit condition exists or tolerance is wasted.

The next step is to determine the LMC clearance hole size, the largest possible clearance hole. The LMC hole size is, essentially, arbitrary. Of course, the clearance hole must be large enough for the fastener plus the stated tolerance, and it cannot be so large that the head of the fastener pulls through the clearance hole.

Some engineers suggest that the clearance hole should not be larger than the largest hole that will fit under the head of the fastener. If a slotted clearance hole, Fig. 8-3A, will fit and function, then surely the .337 diameter hole in Fig. 8-3B will also fit and function. How is the clearance hole diameter in Fig. 8-3B determined? The largest hole that will fit under the head of a fastener is the sum of half of the diameter of the fastener and half of the diameter of the fastener head, or the distance across the flats of the head, as shown in Fig. 8-3C.

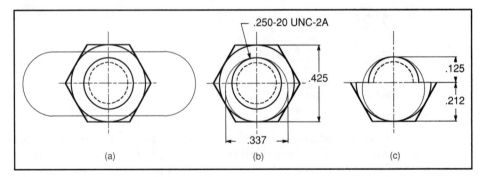

Figure 8-3 Clearance hole size at LMC.

The LMC clearance hole can also be calculated by adding the diameters of the fastener and the fastener head and then dividing the sum by two.

$$H @ \text{LMC} = (F + F\,\text{head})/2$$
$$= (.250 + .425)/2$$
$$= .337$$

This method of selecting the LMC clearance hole size is a rule of thumb that will allow you to compute the largest hole that will fit under the head of the fastener. Engineers may select any size clearance hole that is required, but with the use of the above formula, they can make an informed decision and do not have to blindly depend on an arbitrary clearance hole tolerance chart.

Fixed Fasteners

The fixed fastener is fixed by one or more of the members being fastened. The fasteners in Fig. 8-4 are both fixed; the fastener heads are fixed in their countersunk holes. The fastener, Fig. 8-4B is also fixed in the threaded hole at the

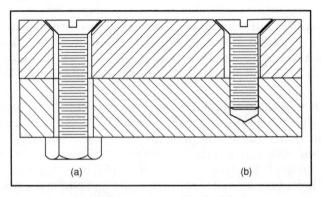

Figure 8-4 A fixed fastener and a double-fixed fastener.

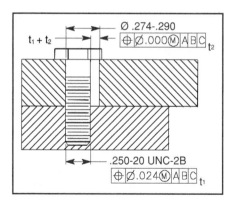

Ø .274-.290

$\oplus$ Ø.000Ⓜ A B C t_2

.250-20 UNC-2B

$\oplus$ Ø.024Ⓜ A B C t_1

$t_1 + t_2$

Figure 8-5 Fixed fastener.

other end of the screw. This screw is considered to be a double-fixed fastener. Double-fixed fasteners should be avoided. It is not always possible to avoid a double-fixed fastener condition where flat-head fasteners are required, but a misaligned double-fixed fastener with a high torque may cause the fastener to fail.

Fixed fasteners are a bit more complicated to calculate than floating fasteners. The formula for fixed fasteners is:

$$t_1 + t_2 = H - F \quad \text{or} \quad H = F + t_1 + t_2$$

Where t_1 is the tolerance for the threaded hole at MMC, t_2 is the tolerance for the clearance hole at MMC, H is the clearance hole diameter at MMC, and F is the fastener diameter at MMC.

This formula is sometimes expressed in terms of $2T$ instead of $t_1 + t_2$; however, $2T$ implies that the tolerances for the threaded and the clearance holes are the same. In most cases, it is desirable to assign more tolerance to the threaded hole than the clearance hole because the threaded hole is usually more difficult to manufacture.

The first step in calculating the tolerance for fixed fasteners is to determine the diameter of the clearance hole at LMC, the largest clearance hole diameter. The engineer might have selected the largest hole that will fit under the head of the quarter-inch fastener, .337, but instead decided to use the more conservative tolerance, .290, shown in Fig. 8-5. The tolerance for both the threaded and the clearance holes must come from the difference between the sizes of the clearance hole and the fastener, the total tolerance available.

$$\text{Total size tolerance} = \text{clearance hole size @ LMC} - \text{fastener}$$
$$= .290 - .250$$
$$= .040$$

Since drilling and taping a hole involves two operations and threading a hole is more problematic than just drilling the hole, it is common practice to assign a larger portion of the tolerance to the threaded hole. In this example, 60 percent of the tolerance is assigned to the threaded hole, and the remaining tolerance applies to the clearance hole.

$$\text{Total tolerance} \times 60\% = .040 \times 60\%$$
$$= .024$$

This position tolerance has a cylindrical tolerance zone .024 in diameter at MMC. Zero positional tolerance is not appropriate for a threaded hole since there is almost no tolerance between threaded features. The tolerance is specified at MMC because there is some movement, however small, between the assembled parts, and some, though small, bonus tolerance is available. Those who are tempted to specify RFS should be aware that costly inspection equipment, a spring thread gage, is required, and a more restrictive tolerance is imposed on the thread. Parts should be toleranced and inspected the way they function in assembly, at MMC.

The fastener, the LMC clearance hole size, and the threaded hole tolerance have all been determined. The clearance hole tolerance and the MMC clearance hole size are yet to be determined. Some individuals like to assign a tolerance of .005 or .010 at MMC to the clearance hole. However, the tolerance at MMC is arbitrary since bonus tolerance is available. Zero tolerance at MMC is as good as any. It has been assigned to the clearance hole in Fig. 8-5 and will be used to calculate the MMC hole diameter.

$$H = F + t_1 + t_2$$
$$= .250 + .024 + .000$$
$$= .274$$

At this point, the engineer may wish to check a drill chart to determine the actual tolerance available. A drill chart and a chart of oversize diameters in drilling are located in the appendix of this text.

TABLE 8-1   Drill Chart

Letter	Fraction	Decimal
	17/64	.266
H	–	.266
I	–	.272
J	–	.277
	9/32	.281
K	–	.281
L	–	.290

The letter L drill would not be used since the drill will probably produce a hole .002 or .003 oversize. If the letter K drill were used and drilled only .002 oversize, the clearance hole tolerance would be

$$Actual\ hole\ size - MMC = tolerance$$
$$.283 - .274 = .009$$

Because of the drill size used, the total tolerance available is not .040 but .033, and the percentage of tolerance assigned to the threaded hole is more than 70 percent of the total tolerance. At this point, the designer may want to increase the hole size or reduce the threaded hole tolerance.

Projected Tolerance Zones

When specifying a threaded hole or a hole for a press fit pin, the orientation of the hole determines the orientation of the mating pin. Although the location and orientation of the hole and the location of the pin will be controlled by the tolerance zone of the hole, the orientation of the pin outside the hole cannot be guaranteed, as shown in Fig. 8-6A. The most convenient way to control the orientation of the pin outside the hole is to project the tolerance zone into the mating part. The tolerance zone must be projected on the same side and at the greatest height of the mating part, as shown in Fig. 8-6B. The height of the tolerance zone is equal to or greater than the thickest mating part or tallest stud or pin after installation. In other words, the tolerance zone height is specified to be at least as tall as the MMC thickness of the mating part or the maximum height of the installed stud or pin. The dimension of the tolerance zone height is specified as a minimum.

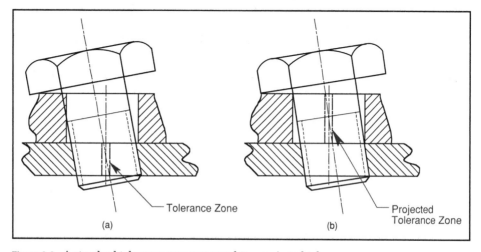

Figure 8-6 A standard tolerance zone compared to a projected tolerance zone.

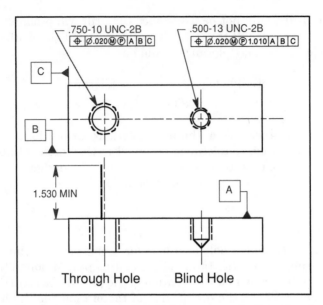

Figure 8-7 Specifying projected tolerance zones for through and blind holes.

When specifying a projected tolerance zone for a *through hole*, place a circle P in the feature control frame after the material condition symbol, and specify both maximum height and direction by drawing and dimensioning a thick chain line next to an extension of the centerline. The chain line is the MMC height of the mating part and located on the side where the mating part assembles. If the mating part is 1.500 ± .030 thick and assembles on top of the plate over the through hole, as shown in Fig. 8-7, the chain line is extended up above the hole and dimensioned with the MMC thickness of the mating part, .530, specified as a minimum.

When specifying a projected tolerance zone for a *blind hole*, place a circle P in the feature control frame after the material condition symbol, and specify the projected MMC height of the mating part after the circle P. If the thickness of the mating part is 1.000 ± .010, then 1.010 is placed in the feature control frame after the circle P, as shown in Fig. 8-7, for blind holes. There is only one direction in which a blind hole can go; therefore, no chain line is drawn.

Multiple Patterns of Features

Where two or more patterns of features are located with basic dimensions, to the same datums features, in the same order of precedence, and at the same material conditions, they are considered to be one composite pattern of features. Even though they are of different sizes and specified at different tolerances, the four patterns of holes in Fig. 8-8 are all located with basic dimensions, to the

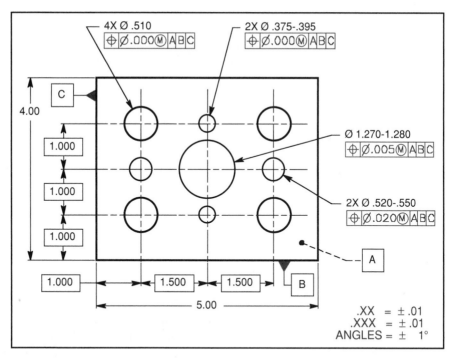

Figure 8-8 Multiple patterns of features located to datum features not subject to size variation (plane surfaces).

same datums features, and in the same order of precedence. (The datums are all plane surfaces; therefore, no material conditions apply.) Consequently, they are to be considered one composite pattern of holes and can be inspected in one setup or with a single gage.

Even though they are of different sizes and specified at different tolerances, the four-hole patterns in Fig. 8-9 are all located with basic dimensions, to the same datum features, in the same order of precedence, and at the same material conditions. The outside diameter, datum feature B, is a size feature specified at RFS. Datum features of size specified at RFS require physical contact between the gaging element and the datum feature. Consequently, the part cannot shift inside a gage or open setup, and the four patterns of holes are to be considered one composite pattern and can be inspected with a single gage or in one inspection setup.

Even though they are of different sizes and specified at different tolerances, the four-hole patterns in Fig. 8-10 are all located with basic dimensions, to the same datum features, in the same order of precedence, and at the same material conditions. The outside diameter, datum feature B, is a size feature specified at MMC. Datum features of size specified at MMC allow a shift tolerance as the datum feature departs from MMC toward LMC. Consequently, a shift tolerance is allowed between datum feature B and the gage; however, if there is no note, the four patterns of holes are to be considered one composite pattern and must

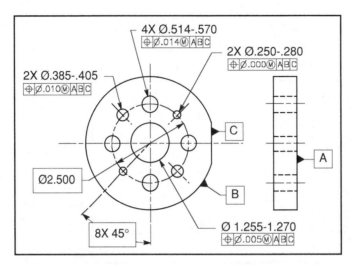

Figure 8-9 Multiple patterns of features located to a datum feature of size specified at RFS.

be inspected in one setup or with a single gage. No matter how the features are specified, as long as they are located with basic dimensions, to the same datums features, in the same order of precedence, and at the same material conditions, the default condition is that patterns of features are to be treated as one composite pattern. If the patterns have no relationship to each other, a note such as "SEP REQT" may be placed under each feature control frame allowing each pattern to be inspected separately. If some patterns are to be

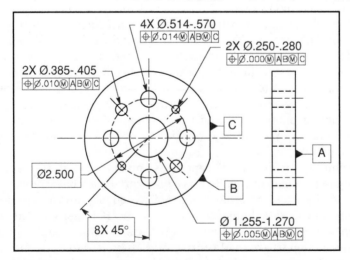

Figure 8-10 Multiple patterns of features located to a datum feature of size specified at MMC.

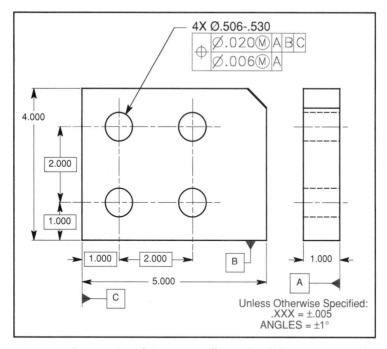

Figure 8-11 A composite tolerance controlling a four-hole pattern to its datums with one tolerance and a feature-to-feature relationship with a smaller tolerance.

inspected separately and some simultaneously, a local note is required to clearly communicate the desired specifications.

Composite Positional Tolerancing

When locating patterns, there are situations where the relationship from feature to feature must be kept to a certain tight tolerance and the relationship between the pattern and its datums is not as critical and may be held to a looser tolerance. These situations often occur when combining technologies that are typically held to different tolerances. For example, composite tolerancing is recommended if a hole pattern on a sheet metal part must be held to a tight tolerance from feature to feature and located from a datum that has several bends between the datum and the pattern requiring a larger tolerance. Also, many industries make machined components that are mounted to a welded frame. The location of the components may be able to float within a tolerance of one-eighth of an inch to the welded frame, but the mounting hole pattern might require a .030 tolerance from feature to feature. Both of these tolerancing arrangements can easily be achieved with composite positional tolerancing.

A composite feature control frame has one position symbol that applies to the two horizontal segments that follow. The upper segment, called the

Figure 8-12 The composite feature control frame.

pattern-locating control, governs the relationship between the datums and the pattern. It acts like any other positional control locating the pattern to datums B and C. Datum A in the upper segment is merely a place holder indicating that datums B and C are secondary and tertiary datums. The lower segment, referred to as the feature-relating control, is a refinement of the upper control and governs the relationship from feature to feature. Each complete horizontal segment in the composite feature control frame must be separately verified, but the lower segment is always a subset of the upper segment. The lower segment is a refinement of the relationship between the features. That is, in Fig. 8-12, the feature-to-feature location tolerance is a cylindrical tolerance zone .006 in diameter at MMC. The primary function of the position control is to control

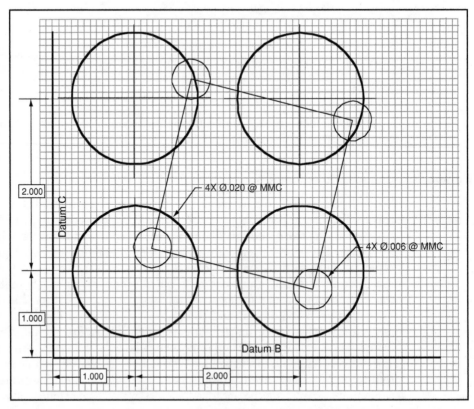

Figure 8-13 A graphic analysis approach to specifying the datum-to-pattern and feature-to-feature tolerance zone relationship for the drawing in Fig. 8-11

location. In addition to controlling location from hole to hole, the Ø.006 tolerance zones are perpendicular to datum A and control the orientation of the features within the same tolerance.

For composite positional tolerancing, there is a requirement and a condition:

- Any datums in the lower segment of the feature control frame are required to repeat the datums in the upper segment. If only one datum is repeated, it would be the primary datum; if two datums were repeated, they would be the primary and secondary datums.

- The condition of datums in the lower segment of the feature control frame is that they only control orientation.

The four Ø.020 cylindrical tolerance zones are centered on their true positions located a basic 1.000 inch and a basic 3.000 inches from datums B and C. These tolerance zones are locked in place. The four Ø.006 cylindrical tolerance zones are centered on their true positions located a basic 2.000 from each other, at right angles to each other, and perpendicular to datum A. These four cylindrical tolerance zones are locked together in a framework. The four Ø.006 cylindrical tolerance zones framework can float, as a pattern, in any direction and rotate about an axis, perpendicular to datum A. A portion of a smaller tolerance zone may fall outside of its respective larger tolerance zone, but that portion is unusable. In other words, the entire feature axis must fall inside both its respective tolerance zones in order to satisfy the requirements specified by the composite feature control frame.

A second datum may be repeated in the lower segment of the feature control frame, as shown in Fig. 8-14. The second datum can only be datum B, and both datums only control the orientation of the smaller tolerance zone framework. Since datum A in the upper segment controls only orientation, i.e., perpendicularly, it is not surprising that datum A in the lower segment is a refinement of perpendicularity to a tighter tolerance. When datum B is included in the lower segment, the Ø.006 cylindrical tolerance zone framework must remain parallel to datum plane B. That means the smaller tolerance zone structure is allowed to translate up and down and left and right but may not rotate about an axis perpendicular to datum A. The tolerance zone framework must remain parallel to datum plane B at all times, as shown in Fig. 8-15.

In a more complex geometry, Fig. 8-16, the four holes are located by the Ø.020 pattern-locating tolerance zones held parallel to and located with a basic dimension from datum plane A, centered on datum axis B, and clocked to datum center plane C. Since datums B and C are size features and specified at MMC, a shift tolerance is allowed. As the datum features depart from MMC toward LMC, the pattern-locating tolerance zones, as a group, can shift with respect to datum axis B and clock about datum axis B as permitted by datum feature C. The pattern location is further refined by the feature-relating control within the Ø.006 cylindrical tolerance zones that may translate in all directions but is held parallel to datum plane A, perpendicular to datum axis B at MMC, and

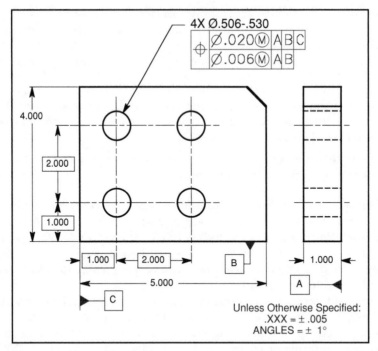

4X Ø.506-.530

$\oplus$ | Ø.020 Ⓜ | A | B | C
Ø.006 Ⓜ | A | B

4.000

2.000

1.000

1.000 | 2.000

B

5.000

C

1.000

A

Unless Otherwise Specified:
.XXX = ± .005
ANGLES = ± 1°

Figure 8-14 A composite positional tolerance with datums A and B repeated in the lower segment of the feature control frame.

parallel to datum center plane C at MMC. As datums B and C depart from MMC toward LMC, a shift tolerance with respect to orientation is allowed for the Ø.006 feature-relating tolerance zones. Each feature axis must fall inside both of its respective tolerance zones.

Two Single-Segment Feature Control Frames

The four-hole pattern in Fig. 8-17 is toleranced with a control called a two single-segment feature control frame. In this case, the lower segment refines the feature-to-feature relationship just as the lower segment of the composite feature control frame does, but the datums behave differently. The lower segment of the two single-segment feature control frame acts just like any other position control. If a datum C were included in the lower segment, the upper segment would be meaningless, and the entire pattern would be controlled to the tighter cylindrical tolerance of Ø.006. In Fig. 8-17, the lower segment of the two single-segment feature control frame refines the feature-to-feature relationship oriented perpendicular to datum A and located to datum B within a Ø.006 cylindrical tolerance. The upper segment allows the feature-relating tolerance zone framework to translate back and forth relative to datum C within a cylindrical tolerance zone .020 in diameter.

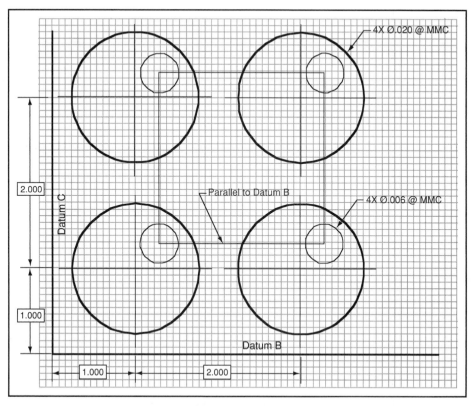

Figure 8-15 A graphic analysis approach to specifying the datum-to-pattern and feature-to-feature tolerance zone relationships with datums A and B repeated in the lower segment of the feature control frame specified in the drawing in Fig. 8-14.

In other words, the smaller tolerance zone framework is locked to datum B by a basic 1.000-inch and cannot move up or down. This control allows only the Ø .006 cylindrical tolerance zone framework to shift back and forth relative to datum C within the larger tolerance zone of Ø .020, as shown in Fig. 8-18.

Nonparallel Holes

The position control is so versatile that it can control a radial pattern of holes at an angle to a primary datum plane. As shown in Fig. 8-19, the radial pattern is dimensioned with a basic 8X 45° to each other and at a basic 8X 30° to datum plane A.

Counterbored Holes

Counterbores that have the same location tolerance as their respective holes are specified by indicating the hole callout and the counterbore callout followed

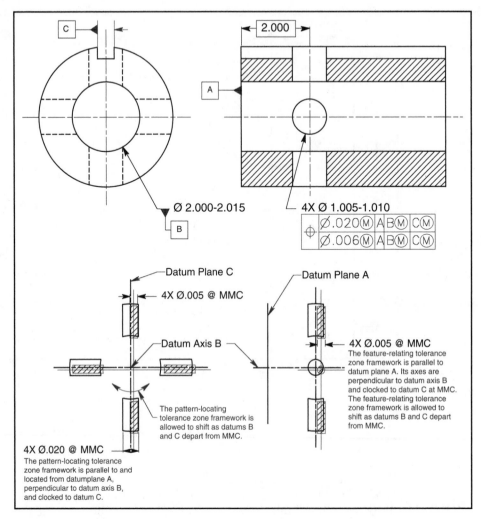

Figure 8-16 A composite positional tolerance with three datums in the upper and lower segments.

by the geometric tolerance for both. The counterbore callout includes the counterbore symbol, the diameter symbol, the size dimension, and the tolerance. The depth is specified using the depth symbol followed by the depth dimension and tolerance. The feature control frame locating both the hole and counterbore patterns is placed below. The complete callout is shown in Fig. 8-20.

Counterbores with a larger location tolerance than their respective holes, however, are specified by separating the hole callout from the counterbore callout. After specifying the hole pattern callout and its geometric tolerance, the complete counterbore callout is stated followed by its larger geometric tolerance, as shown in Fig. 8-21. Note that "4X" is repeated before the counterbore callout.

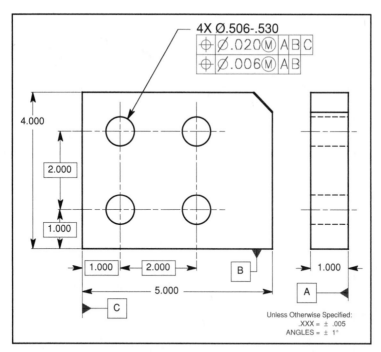

4X Ø.506-.530

⊕ | Ø.020Ⓜ | A | B | C

⊕ | Ø.006Ⓜ | A | B

4.000

2.000

1.000

1.000   2.000

5.000

B

C

A

1.000

Unless Otherwise Specified:
.XXX = ± .005
ANGLES = ± 1°

Figure 8-17 A two single-segment feature control frame is used to control a four-hole pattern to datums A and B with a Ø.006 cylindrical tolerance and to datum C with a Ø.020 cylindrical tolerance.

Finally, counterbores with a smaller location tolerance than their respective holes are toleranced by first specifying the hole callout followed by the geometric tolerance. Then, each counterbore is located to its respective hole by identifying one of the holes as datum C (including the note "4X INDIVIDUALLY" next to the datum feature symbol) and tolerancing the counterbore relative to datum C (again including the note "4X INDIVIDUALLY" beneath the feature control frame, as shown in Fig. 8-22.)

Noncircular Features at MMC

Elongated holes are dimensioned from specified datums to their center planes with basic dimensions. The feature control frames are associated with the size dimensions in each direction. If only one tolerance applies in both directions, one feature control frame may be attached to the elongated hole with a leader not associated with the size dimension. No diameter symbol precedes the tolerance in the feature control frame since the tolerance zone is not a cylinder. The note "BOUNDARY" is placed beneath each feature control frame. Each elongated hole must be within its size limits, and no element of the feature surface may fall inside its virtual condition boundary. The virtual condition boundary is the exact shape of the elongated hole and equal in size to its virtual condition.

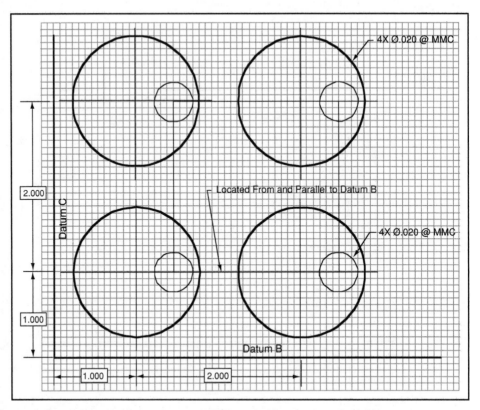

4X Ø.020 @ MMC

Located From and Parallel to Datum B

4X Ø.020 @ MMC

2.000

Datum C

1.000

Datum B

1.000

2.000

Figure 8-18 The two single-segment control allows the pattern of smaller tolerance zones to move back and forth within the larger tolerance zones but does not allow movement up and down.

Figure 8-23 shows elongated holes that are .50 by 1.00 with a size tolerance of ±.010. The boundary is equal to the MMC minus the geometric tolerance, i.e., .490 − .020 = .470 for the width, and .990 − .060 = .930 for the length.

Symmetrical Features at MMC

A size feature may be located symmetrically to a datum feature of size and toleranced with a position control associated with the size dimension of the feature being controlled. No diameter symbol precedes the tolerance in the feature control frame since the tolerance zone is not a cylinder.

The position tolerance zone to control symmetry consists of two parallel planes evenly disposed about the center plane of the datum feature and separated by the geometric tolerance. The drawing in Fig. 8-24 has a slot symmetrically controlled to datums A and B. Since datum A is the primary datum, the tolerance zone is first perpendicular to datum A and then located symmetrically to datum B at MMC. The circle M symbol after the geometric tolerance provides

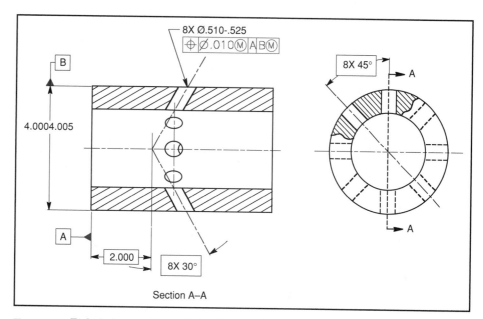

Figure 8-19 Eight holes specified radially about a cylinder and at a 30° angle to datum plane A.

the opportunity for a bonus tolerance as the feature departs from MMC toward LMC in the exact amount of such departure. The circle M symbol after the datum provides the opportunity for a shift tolerance as the datum feature departs from MMC toward LMC in the exact amount of such departure.

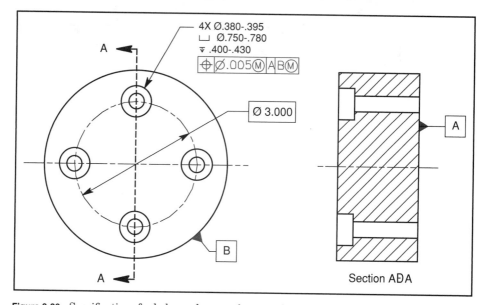

Figure 8-20 Specifications for holes and counterbores with the same tolerances for both.

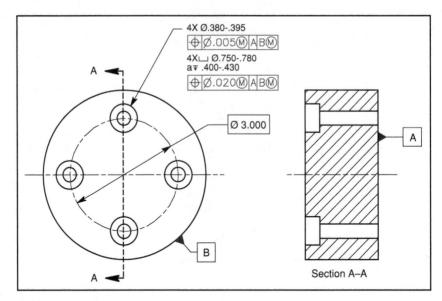

Figure 8-21 Specifications for holes and counterbores with a larger tolerance for the counterbores.

If the datum feature is produced at 4.002 at MMC and the slot is produced at 2.000 also at MMC, then the position tolerance is .010 as stated in the feature control frame. If the datum feature remains the same size but the slot becomes larger, a bonus tolerance is available. If the slot remains the same size but the datum feature becomes smaller, a shift tolerance is available. Of course, as

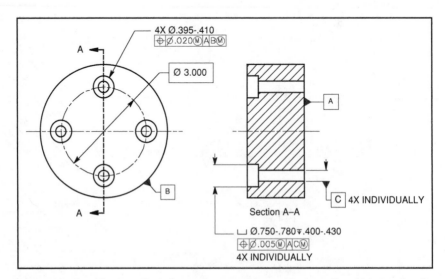

Figure 8-22 Specifications for holes and counterbores with a smaller tolerance for the counterbores.

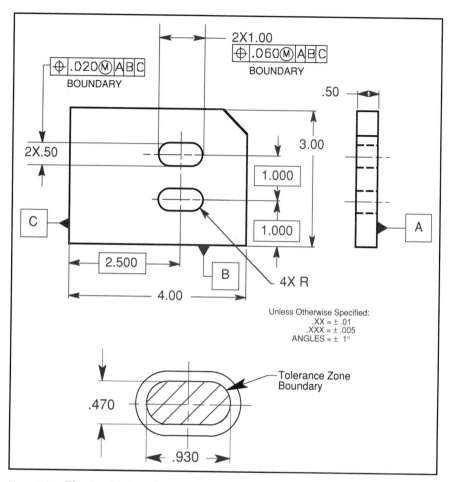

Figure 8-23 Elongated holes toleranced with the position control in the length and width directions.

they both change size from MMC toward LMC, the slot gains bonus tolerance and shift tolerance in addition to the .010 positional tolerance specified in the feature control frame. The part in Fig. 8-24 is a special case for shift tolerance. Where there is only one feature being controlled to the datum feature, the entire shift tolerance is applied to the slot, a single feature. For the more general condition where a pattern of features is controlled to a datum feature of size, the shift tolerance does not apply to each individual feature but applies to the entire pattern of features as a group.

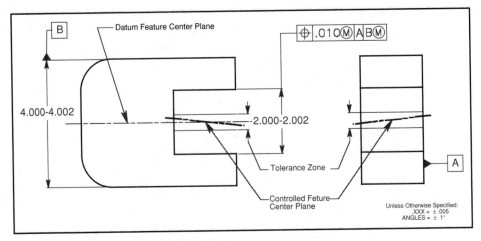

Figure 8-24 A position tolerance used to control the symmetry between size features.

TABLE 8-2 As the Sizes of the Feature and the Datum Feature Depart from MMC toward LMC, the Feature Gains Positional Tolerance

	Size of feature		
Size of datum	**2.000**	**2.001**	**2.002**
4.002	.010	.011	.012
4.001	.011	.012	.013
4.000	.012	.013	.014

Summary

- The floating fastener formula is: $T = H - F$ or $H = F + T$
- The fixed fastener formula is: $t_1 + t_2 = H - F$ or $H = F + t_1 + t_2$
- The LMC clearance hole formula is: $H @ \text{LMC} = (F + F\,\text{head})/2$
- Projected tolerance zone: The most convenient way to control the orientation of a pin outside a threaded or press fit hole is to project the tolerance zone into the mating part.
- Multiple patterns of features: No matter how the features are specified, the default condition is that patterns of features are to be treated as one composite pattern as long as they are located
  - With basic dimensions
  - To the same datums features
  - In the same order of precedence
  - At the same material conditions

If the patterns are specified at MMC and have no relationship to one another, a note such as "SEP REQT" may be placed under each feature control frame allowing each pattern to be inspected separately.

- Composite positional tolerancing: A composite tolerance controls a pattern of features to its datums with one tolerance and a feature-to-feature relationship with a smaller tolerance.

- When datum B is included in the lower segment of a composite feature control frame, the smaller tolerance zone framework must remain parallel to datum B.

- The lower segment of a two single-segment feature control frame acts just like any other position control. The lower segment refines the feature-to-feature tolerance zone framework by orienting it to the primary datum and locating it to the secondary datum with basic dimensions.

- Nonparallel holes: The position control is so versatile that it can control patterns of nonparallel holes at a basic angle to a principle plane.

- Counterbored holes: Counterbores can be toleranced with the same tolerance, more tolerance, or less tolerance than their respective holes.

- Noncircular features at MMC: Elongated holes are dimensioned and toleranced in both directions. The feature control frames do not have cylindrical tolerance zones but have a note BOUNDARY placed beneath them.

- Symmetrical features at MMC: A size feature may be located symmetrically to a datum feature of size and toleranced with a position control associated with the size dimension of the feature being controlled.

Chapter Review

1. The floating fastener formula is _____

   _____

2. $T =$ _____ $H =$ _____ $F =$ _____

3. The LMC clearance hole can be calculated by _____.

4. The fixed fastener is fixed by one or more of the _____.

5. A fastener fixed at its head in a countersunk hole and in a threaded hole at the other end is called what? _____

6. The formula for fixed fasteners is

   _____

7. The tolerance for both the threaded hole and the clearance hole must come from the difference between the size of the clearance hole and the size of the _____.

8. Total possible tolerance equals clearance hole size @ LMC minus _____.

   _____

9. It is common to assign a larger portion of the tolerance to the _____ hole.

10. As much as 60% of the tolerance may be assigned to the _____ hole.

11. When specifying a threaded hole or a hole for a press fit pin, the orientation of the _____ determines the orientation of the mating pin.

12. The most convenient way to control the orientation of the pin outside the hole is to _____ the tolerance zone into the mating part.

13. The height of the projected tolerance zone is equal to or greater than the thickest _____ or tallest _____ after installation.

14. The dimension of the projected tolerance zone height is specified as a

_____.

15. Two or more patterns of features are considered to be one composite pattern if they _____

_____

_____

_____.

16. Datum features of size specified at RFS require _____ between the gagging element and the datum feature.

17. If the patterns of features have no relationship to one another, a note such as _____ may be placed under each feature control frame allowing each pattern to be inspected separately.

18. Composite tolerancing allows the relationship from _____ to be kept to a tight tolerance and the relationship between the _____ to be controlled to a looser tolerance.

19. A composite positional feature control frame has one _____ symbol that applies to the two horizontal _____ that follow.

20. The upper segment of a composite feature control frame, called the _____ control, governs the relationship between the datums and the _____.

21. The lower segment of a composite feature control frame is called the _____ control; it governs the relationship from _____.

22. The primary function of the position control is to control _____.

23. There is a requirement and a condition for the datums in the lower segment of the composite positional tolerancing feature control frame. They:

_____

_____

(Assume plane surface datums for question numbers 24 and 25.)

24. When the secondary datum is included in the lower segment of a composite feature control frame, the tolerance zone framework must remain _____ to the secondary datum plane.

25. The lower segment of a two single-segment feature control frame refines the feature-to-feature relationship _____ to the primary datum plane and _____ to the secondary datum plane.

26. Counterbores that have the same location tolerance as their respective holes are specified by indicating the _____

_____.

27. Counterbores that have a larger location tolerance than their respective holes are specified by _____.

28. When tolerancing elongated holes, no _____ precedes the tolerance in the feature control frame since the tolerance zone is not a _____. The note _____ is placed beneath each feature control frame.

29. The virtual condition boundary is the _____ of the elongated hole and equal in size to its _____.

30. A _____ may be located symmetrically to a datum feature of size and toleranced with a _____ associated with the size dimension of the feature being controlled.

Problems

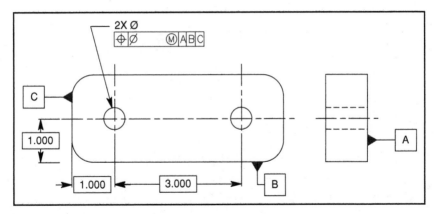

Figure 8-25 Floating fastener drawing: Problems 1 through 4.

1. Specify the MMC and LMC clearance hole sizes for #10 (Ø.190) socket head cap screws.

    | ⊕ | Ø.030Ⓜ | A | B | C | | ⊕ | Ø.010Ⓜ | A | B | C | | ⊕ | Ø.000Ⓜ | A | B | C |

2. If the actual size of the clearance holes in problem 1 is Ø.230, calculate the total positional tolerance for each callout.

Actual size	.230	.230	.230
MMC			
Bonus			
Geometric tolerance			
Total tolerance			

3. Specify the MMC and LMC clearance hole sizes for 3/8 (Ø .375) hex head bolts.

$\oplus$ | Ø.025Ⓜ|A|B|C $\oplus$ | Ø.015Ⓜ|A|B|C $\oplus$ | Ø.000Ⓜ|A|B|C

4. If the clearance holes in problem 3 actually measure Ø .440, calculate the total positional tolerance for each callout.

Actual Size	.440	.440	.440
MMC			
Bonus			
Geometric Tolerance			
Total Tolerance			

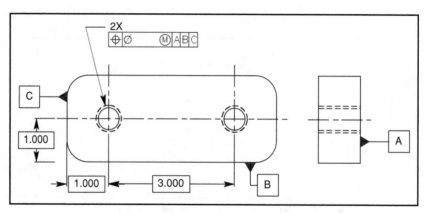

Figure 8-26   Fixed fastener drawing: Problems 5 through 8.

5. Specify the MMC and LMC clearance hole sizes for #8 (Ø .164) socket head cap screws.

2X Ø .164 (#8)-32 UNF-2B   2X Ø .164 (#8)-32 UNF-2B   2X Ø .164 (#8)-32 UNF-2B

$\oplus$ | Ø.025Ⓜ|A|B|C $\oplus$ | Ø.025Ⓜ|A|B|C $\oplus$ | Ø.025Ⓜ|A|B|C

$\oplus$ | Ø.010Ⓜ|A|B|C $\oplus$ | Ø.005Ⓜ|A|B|C $\oplus$ | Ø.000Ⓜ|A|B|C

6. If the clearance holes in problem 5 actually measure Ø .205, calculate the total positional tolerance for each callout.

Actual Size	.205	.205	.205
MMC			
Bonus			
Geometric Tolerance			
Total Tolerance			

7. Specify the MMC and LMC clearance hole sizes for the 1/2 hex head bolts.

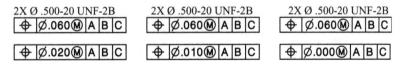

2X Ø .500-20 UNF-2B	2X Ø .500-20 UNF-2B	2X Ø .500-20 UNF-2B
⊕ Ø.060Ⓜ A B C	⊕ Ø.060Ⓜ A B C	⊕ Ø.060Ⓜ A B C
⊕ Ø.020Ⓜ A B C	⊕ Ø.010Ⓜ A B C	⊕ Ø.000Ⓜ A B C

8. If the clearance holes in problem 5 actually measure Ø .585, calculate the total positional tolerance for each callout.

Actual Size	.585	.585	.585
MMC			
Bonus			
Geometric Tolerance			
Total Tolerance			

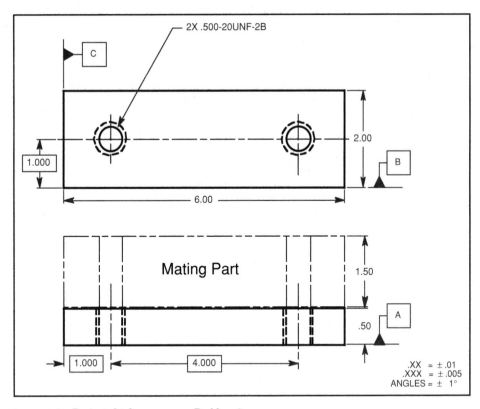

Figure 8-27 Projected tolerance zone: Problem 9.

9. Complete the drawing in Fig. 8-27. Specify a Ø .040 tolerance at MMC with the appropriate projected tolerance.

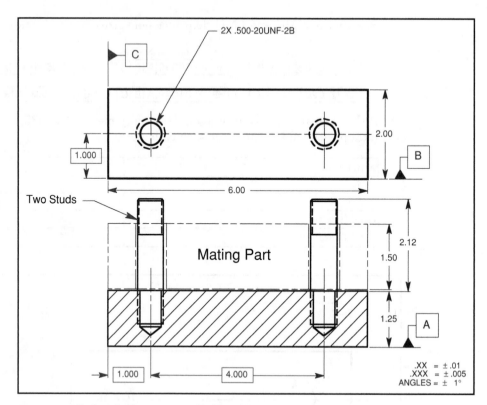

Figure 8-28 Projected tolerance zone: Problem 10.

10. Complete the drawing in Fig. 8-28. Specify a Ø .050 tolerance at MMC with the appropriate projected tolerance.

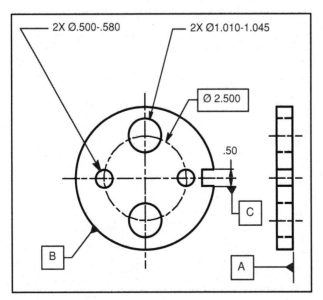

2X Ø.500-.580   2X Ø1.010-1.045

Ø 2.500

.50

C

B

A

Figure 8-29   Multiple patterns of features: Problems 11 through 13.

11. Position the small holes with a Ø.000 tolerance at MMC and the large holes with Ø.010 tolerance at MMC; locate them to the same datums and in the same order of precedence. Use MMC wherever possible.

12. Must the hole patterns be inspected in the same setup or in the same gage? Explain. _____
_____
_____
_____
_____

13. Can the requirement be changed, how?
_____
_____
_____

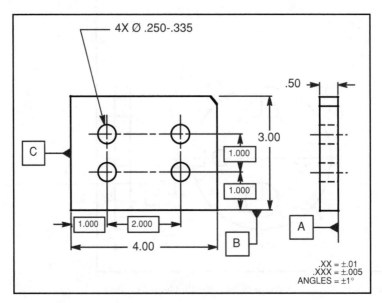

Figure 8-30   Composite tolerancing: Problems 14 and 15.

14. The pattern of clearance holes in the part in Fig. 8-30 must be located within a cylindrical tolerance zone of Ø .060 at MMC to the datums specified. The plate is designed to be assembled to the mating part with 1/4-inch bolts as floating fasteners. Complete the drawing.

15. It has been determined that the hole pattern in Fig. 8-30 is required to remain parallel, within the smaller tolerance, to datum B. Draw the feature control frame that will satisfy this requirement.

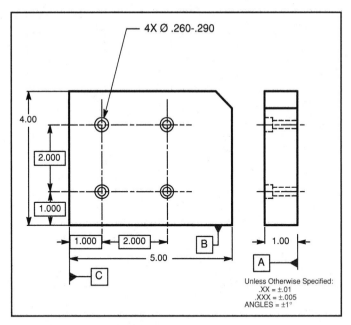

4X Ø .260-.290

4.00

2.000

1.000

1.000   2.000   B

5.00

C

1.00

A

Unless Otherwise Specified:
.XX = ±.01
.XXX = ±.005
ANGLES = ±1°

Figure 8-31   Counterbore: Problems 16 and 17.

16. Tolerance the holes and counterbores in Fig. 8-31 for four Ø .250 socket head cap screws. The counterbores are Ø .422 ± .010, the depth is .395 ± .010, and the geometric tolerance is .010 at MMC.

17. If the geometric tolerance for just the counterbores in Fig. 8-31 can be loosened to .020 at MMC instead of .010, draw the entire callout.

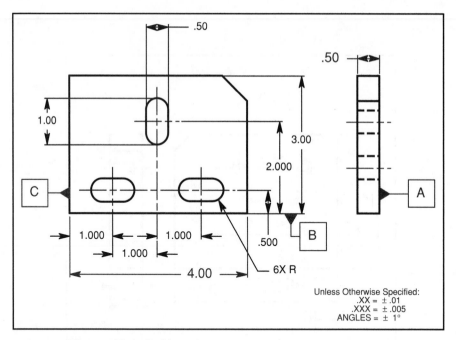

Figure 8-32 Elongated hole: Problem 18.

18. Specify a geometric tolerance of .040 at MMC in the .500-inch direction and .060 at MMC in the 1.000-inch direction for the elongated holes in Fig. 8-32.

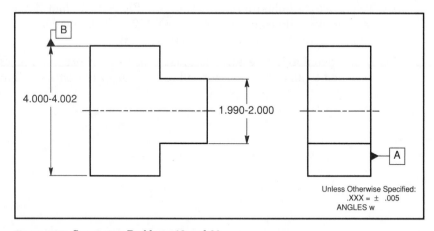

Figure 8-33 Symmetry: Problems 19 and 20.

19. Control the 2.000-inch feature in Fig. 8-33 symmetrical with the 4.000-inch feature within a tolerance of .020 at MMC to the datum indicated. Use MMC wherever possible.

20. If the controlled feature in Fig. 8-33 happened to be produced at 1.995 and the datum feature produced at 4.000, what would the total positional tolerance be? _____

Position, Coaxiality

One of the most common drawing errors is the failure to specify coaxiality tolerance. Many practitioners think coaxiality tolerance is unnecessary or are not even aware that coaxiality must be toleranced. The position tolerance used to control coaxiality will be discussed in this chapter.

Chapter Objectives

After completing this chapter, you will be able to

- *Explain* the difference between position, runout, and concentricity
- *Specify* position tolerance for coaxiality.
- *Specify* coaxiality on a material condition basis
- *Specify* composite positional control of coaxial features
- *Tolerance* a plug and socket

Definition

Coaxiality is that condition where the axes of two or more surfaces of revolution are coincident.

Many engineers produce drawings similar to the one in Fig. 9-1, showing two or more cylinders on the same axis. This is an incomplete drawing because there is no coaxiality tolerance. It is a misconception that centerlines or the tolerance block control the coaxiality between two cylinders. The centerlines indicate that the cylinders are dimensioned to the same axis. In Fig. 9-1, the distance between the axes of the Ø 1.000-inch and Ø 2.000-inch cylinders is zero. Of course, zero dimensions are implied and never placed on drawings. Even though the dimension is implied, the tolerance is not; there is no tolerance

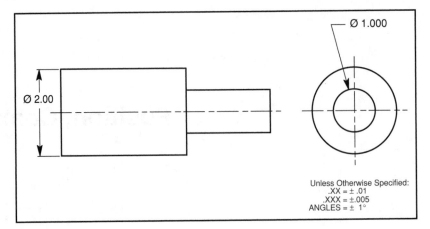

Figure 9-1 Definition—two surfaces of revolution on the same axis.

indicating how far out of coaxiality the axes of an acceptable part may be. Many practitioners erroneously think title block tolerances control coaxiality. They do not. See Rule #1 in Chapter 4, "the relationship between individual features," for a more complete discussion of the tolerance between individual features.

There are other methods of controlling coaxiality such as a note or a dimension and tolerance between diameters, but a geometric tolerance, such as the one in Fig. 9-2, is preferable. The position control is the appropriate tolerance for coaxial surfaces of revolution that are cylindrical and require the maximum material condition (MMC) or the least material condition (LMC). The position control provides the most tolerancing flexibility.

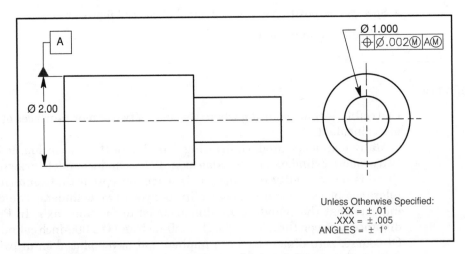

Figure 9-2 Two surfaces of revolution toleranced for coaxiality.

Comparison Between Position, Runout, and Concentricity

The standard specifically states, "The amount of permissible variation from coaxiality may be expressed by a position tolerance or a runout tolerance." In general, a position control is used when parts are mated in a static assembly, and runout is specified for high-speed rotating assemblies.

Many people erroneously specify a concentricity tolerance for the control of coaxiality, perhaps because they use the terms coaxial and concentric interchangeably. Coaxial means that two or more features have the same axis. Concentric means that two or more plane geometric figures have the same center. A concentricity tolerance is the control of all median points of a figure of revolution within a cylindrical tolerance zone. Although concentricity is not strictly a coaxiality control, in effect, it does control coaxiality. However, the concentricity control requires an expensive inspection process and is appropriate in only a few unique applications where precise balance is required.

TABLE 9-1   A Comparison Between Position, Runout, and Concentricity

Characteristic symbol	Tolerance zone	Material condition	Surface error
⊕	⌀	Ⓜ Ⓛ	Includes
↗ & ↗↗	Two concentric circles/cylinders	None	Includes
◎	⌀	None	Independent

Specifying Coaxiality at MMC

Coaxiality is specified by associating a feature control frame with the size dimension of the feature being controlled. A cylindrical tolerance zone is used to control the axis of the toleranced feature. Both the tolerance and the datum(s) may apply at maximum material condition, least material condition, or regardless of feature size. At least one datum must be specified in the feature control frame.

When a coaxiality tolerance and a datum feature of size are specified at MMC, bonus and shift tolerances are available in the exact amount of such departures from MMC. The circle M symbol after the geometric tolerance provides the opportunity for a bonus tolerance as the feature departs from MMC toward LMC. The circle M symbol after the datum provides the opportunity for a shift tolerance as the datum feature departs from MMC toward LMC. If the datum feature is produced at 4.002, MMC, and the ⌀ 2.000 cylinder is produced at 2.003, also MMC, then the position tolerance is .005 as stated in the feature

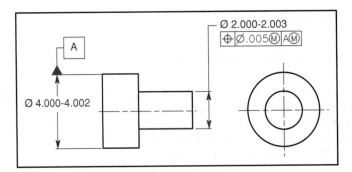

Figure 9-3 Specifying coaxiality at MMC to a datum at MMC.

control frame. If the datum feature remains the same size but the Ø 2.000 cylinder is produced smaller, bonus tolerance becomes available in the exact amount of such departure from MMC. If the Ø 2.000 cylinder remains the same size but the datum feature is produced smaller, a shift tolerance is available in the exact amount of such departure from MMC. Of course, as they both change size from MMC toward LMC, the Ø 2.000 cylinder gains bonus tolerance and shift tolerance in addition to the .005 positional tolerance specified in the feature control frame. The part in Fig. 9-3 is a special case for shift tolerance. Where there is only one feature being controlled to the datum feature, the entire shift tolerance is applied to the Ø 2.000 cylinder, a single feature. For the more general condition where a pattern of features is controlled to a datum feature of size, the shift tolerance does not apply to each individual feature. The shift tolerance applies to the entire pattern of features as a group.

TABLE 9-2 As the Size of the Feature and the Size of the Datum Feature Depart from MMC Toward LMC, the Feature Gains Positional Tolerance

	Size of feature			
Size of datum	2.003	2.002	2.001	2.000
4.002	.005	.006	.007	.008
4.001	.006	.007	.008	.009
4.000	.007	.008	.009	.010

Composite Positional Control of Coaxial Features

A composite positional tolerance may be applied to a pattern of coaxial features such as those in Fig. 9-4. The upper segment of the feature control frame controls the location of the hole pattern to datums A and B. The lower segment of the feature control frame controls the coaxiality of the holes to one another within the tighter tolerance. The smaller tolerance zone may float up and down, back and forth, and at any angle to datums A and B. Portions of the smaller tolerance zone may fall outside the larger tolerance zone, but these portions are unusable. The axes of the holes must fall inside both of their respective tolerance zones at the same time.

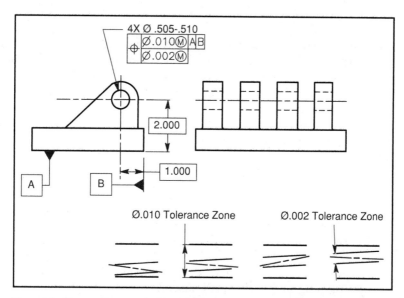

Figure 9-4 Composite control of coaxial features.

Datums in the lower segment of a composite feature control frame only control orientation and must repeat the datums in the upper segment. Since the datums in Fig. 9-5 are repeated in the lower segment of the feature control frame, the smaller tolerance zone may float up and down, back and forth, but must remain parallel to datums A and B. The axes of the holes must fall inside both of their respective tolerance zones at the same time.

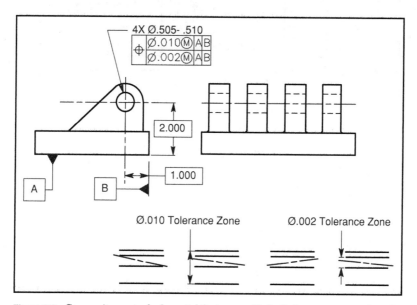

Figure 9-5 Composite control of coaxial features with both datums repeated in the lower segment.

Tolerancing a Plug and Socket

When an external, cylindrical step part is required to assemble inside an internal mating part, diameters, such as the datum features, are dimensioned to mate. Some designers feel strongly that internal and external features should not have the same maximum material conditions. They are concerned that a line fit will result. However, it is extremely unlikely that both parts would be manufactured at MMC. If additional clearance is required, tolerance accordingly. Once the datums have been dimensioned, tolerance the step features to their virtual conditions.

	Plug	Socket
Maximum material condition	.500	.505
Geometric tolerance	+.000	−.005
Virtual condition	.500	.500

A mating plug and socket will assemble every time if they are designed to their virtual conditions as shown in Fig. 9-6.

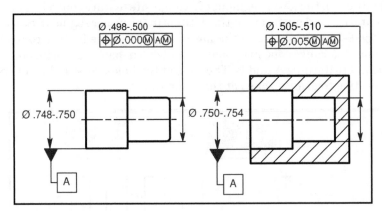

Figure 9-6 Plug and socket.

Summary

- Coaxiality is that condition where the axes of two or more surfaces of revolution are coincident.

- The amount of permissible variation from coaxiality may be expressed by a position tolerance or a runout tolerance. Although concentricity is not strictly a coaxiality control, in effect, it does control coaxiality. However, the concentricity control requires an expensive inspection process and is appropriate in only a few unique applications where precise balance is required.

- A coaxiality control has a cylindrical tolerance zone and may apply at MMC, LMC, or RFS.

- When a coaxiality tolerance and a datum feature of size are specified at MMC, bonus and shift tolerances are available in the exact amount of such departures from MMC.

- A composite positional tolerance may be applied to a pattern of coaxial features.

- A mating plug and socket will assemble every time if they are designed to their virtual conditions.

Chapter Review

1. Coaxiality is that condition where the axes of two or more surfaces of revolution are _____.

2. It is a misconception that centerlines or the tolerance block control the _____ between two cylinders.

3. The _____ control is the appropriate tolerance for coaxial surfaces of revolution that are cylindrical and require MMC or LMC.

4. A _____ tolerance zone is used to control the axis of a feature toleranced with a position or a concentricity control.

5. For position, both the tolerance and the datum(s) may apply at what material conditions? _____

6. When a coaxiality tolerance and a datum feature of size are specified at maximum material conditions, _____ tolerances are available in the exact amount of the departures from MMC toward LMC.

7. The upper segment of a composite feature control frame controls the location of the hole pattern to _____.

8. The lower segment of a composite feature control frame controls the coaxiality of holes to _____.

9. The smaller tolerance zone of a composite feature control frame with no datums may float _____.

10. A mating plug and socket will assemble every time if they are designed to their _____.

Problems

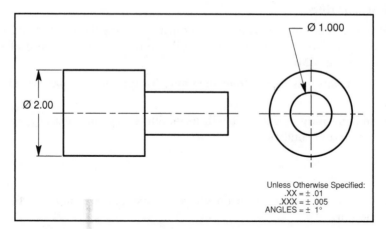

Ø 1.000

Ø 2.00

Unless Otherwise Specified:
.XX = ± .01
.XXX = ± .005
ANGLES = ± 1°

Figure 9-7   Specify coaxiality: Problem 1–3.

1. What controls the coaxiality of the two cylinders in the drawing in Fig.9-7?

2. In the drawing in Fig. 9-7, specify coaxiality tolerance to control the Ø 1.000 feature within a cylindrical tolerance zone of .004 to the Ø 2.000 feature. Use MMC wherever possible.

3. Now that you have added the feature control frame to the drawing in Fig. 9-7, if the larger diameter is produced at 2.000 and the smaller diameter is produced at 1.000, how much total coaxiality tolerance applies?

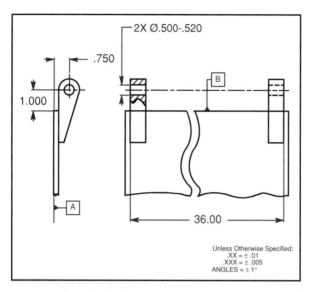

Figure 9-8 Specify coaxiality: Problem 4.

4. In the drawing in Fig. 9-8, locate the two holes in the hinge brackets within .030 at MMC to the datums indicated, and specify their coaxiality to each other. They must be able to accept a Ø .500-inch hinge pin. Specify MMC wherever possible.

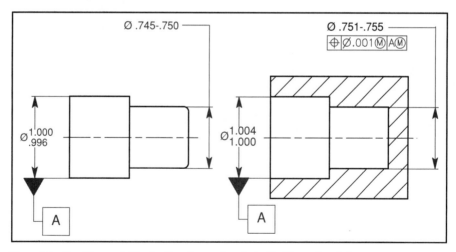

Figure 9-9 Specify coaxiality for the plug and socket—Problems 5 and 6.

5. Control the coaxiality of both parts shown in Fig. 9-9 so that they will always assemble.

6. Draw and dimension the tolerance zones at MMC on the drawing in Fig. 9-9.

10

Concentricity and Symmetry

Both concentricity and symmetry controls are reserved for a few unique tolerancing applications. The controls employ the same tolerancing concept but apply to different geometries. Concentricity controls features constructed about an axis, and symmetry controls features constructed about a center plane. Concentricity and symmetry both locate features by controlling their center points within a specified tolerance zone. They are typically used when it is important to accurately balance the mass of a part about its axis or center plane.

Chapter Objectives

After completing this chapter, you will be able to

- *Define* concentricity and symmetry
- *Specify* concentricity and symmetry on drawings
- *Describe* the inspection process of concentricity and symmetry
- *Explain* applications of concentricity and symmetry

Concentricity

Definition

Concentricity is that condition where the median points of all diametrically opposed points of a surface of revolution are congruent with the axis (or center point) of a datum feature. Concentricity applies to correspondingly located points of two or more radically disposed features, such as the flats on a regular hexagon, or opposing lobes on features such as an ellipse.

Specifying concentricity

Concentricity is a location control. It has a cylindrical-shaped tolerance zone that is coaxial with the datum axis. Concentricity tolerance applies only on a

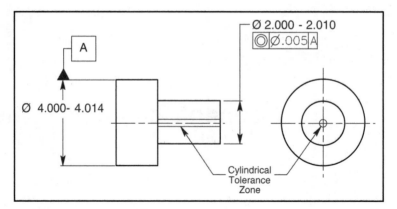

Figure 10-1   Concentricity has a cylindrical tolerance zone and applies at RFS.

regardless of feature size (RFS) basis; it must have at least one datum that also applies only at RFS. The feature control frame is usually placed beneath the size dimension or attached to an extension of the dimension line. The concentricity tolerance has no relationship to the size of the feature being controlled and may be either larger or smaller than the size tolerance. If the concentricity tolerance is specified to control the location of a sphere, the tolerance zone is spherical and its center point is basically located from the datum feature(s).

Interpretation

Concentricity controls all median points of all diametrically opposed points on the surface of the toleranced feature. The aggregate of all median points, sometimes described as a "cloud of median points," must lie within a cylindrical tolerance zone whose axis is coincident with the axis of the datum feature. The concentricity tolerance is independent of both size and form. Differential measurement excludes size, shape, and form while controlling the median points of the feature. The feature control frame in Fig. 10-2 specifies a cylindrical

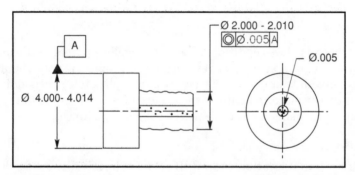

Figure 10-2   A concentricity tolerance locating a coaxial feature.

tolerance zone .005 in diameter and coaxial with the datum axis. Differential measurements are taken along and around the toleranced feature to determine the location of its median points. If all median points fall inside the tolerance zone, the feature is in tolerance.

Inspection

Concentricity can be inspected, for acceptance only, by placing a dial indicator on the toleranced surface of revolution and rotating the part about the datum axis. If the full indicator movement (FIM) on the dial indicator does not exceed the specified tolerance, the feature is acceptable. This technique is a simple first check that will accept parts but will not reject them, and it can be used only on surfaces of revolution. Features such as regular polygons and ellipses must be inspected using the traditional method of differential measurements. If the measurement does exceed the FIM, the part is not necessarily out of tolerance. To reject a part with a concentricity tolerance, the datum is placed in a chucking device that will rotate the part about its datum axis. A point on the surface of the toleranced feature is measured with a dial indicator. The part is then rotated 180° so the diametrically opposed point can be measured. The difference between the measurements of the two points determines the location of the median point. This process is repeated a predetermined number of times. If all median points fall within the tolerance zone, the feature is in tolerance. The size and form, Rule # 1, are measured separately.

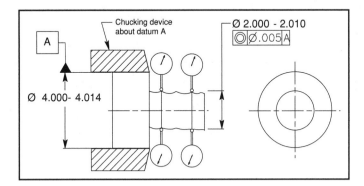

Figure 10-3 Inspecting a part with a concentricity tolerance.

Applications of concentricity

The concentricity tolerance is often used to accurately control balance for high-speed rotating parts. Runout also controls balance, but it controls form and surface imperfections at the same time. Runout is relatively easy and inexpensive to inspect, but manufacturing is more difficult and costly. Concentricity is time-consuming and expensive to inspect but less expensive to manufacture since it is not as rigorous a requirement as runout. Concentricity is appropriately used for large, expensive parts that must have a small coaxial tolerance for balance but need not have the same small tolerance for form and surface imperfections. Concentricity is also used to control the coaxiality of noncircular features such as regular polygons and ellipses.

Symmetry

Definition

Symmetry is that condition where the median points of all opposed or correspondingly located points of two or more feature surfaces are congruent with the axis or center plane of a datum feature.

Specifying symmetry

Symmetry is a location control. It has a tolerance zone that consists of two parallel planes evenly disposed about the center plane or axis of the datum feature. Symmetry tolerance applies only at RFS; it must have at least one datum that also applies only at RFS. A feature control frame is usually placed beneath the size dimension or attached to an extension of the dimension line. The symmetry tolerance has no relationship to the size of the feature being controlled and may be either larger or smaller than the size tolerance.

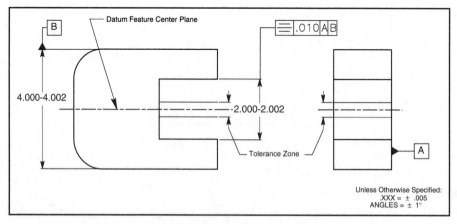

Figure 10-4 The symmetry tolerance zone consists of two parallel planes.

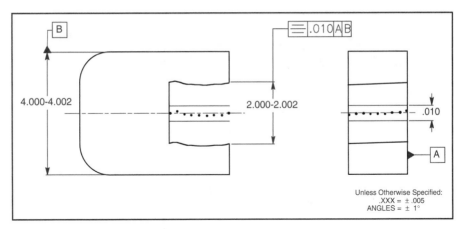

Figure 10-5 A symmetry tolerance locating a symmetrical feature.

Interpretation

Symmetry controls the median points of all opposed or correspondingly located points of two or more surfaces. The aggregate of all median points, sometimes described as a "cloud of median points," must lie within a tolerance zone defined by two parallel planes equally disposed about the center plane of the datum feature, i.e., half of the tolerance is on one side of the center plane, and half is on the other side. The symmetry tolerance is independent of both size and form. Differential measurement excludes size, shape, and form while controlling the median points of the feature. The feature control frame in Fig. 10-5 specifies a tolerance zone consisting of two parallel planes .010 apart, perpendicular to datum plane A, and equally disposed about datum plane B. Differential measurements are taken between the two surfaces to determine the location of the median points. If all median points fall inside the tolerance zone, the feature is in tolerance.

Inspection

A simple method of measuring symmetry is shown in Fig. 10-6. This method can be used only if the datum surfaces are parallel compared to the symmetry tolerance. In this example, one of the datum surfaces is placed on the surface plate. A dial indicator is used to measure a number of points on the surface of the slot. These measurements are recorded. The part is turned over, and the process is repeated. The measurements are compared to determine the location of the median points and whether or not the feature is in tolerance. The size and form, Rule # 1, are measured separately.

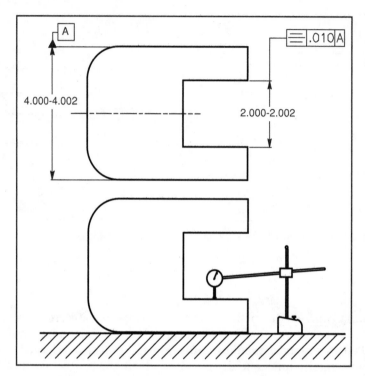

Figure 10-6 Inspecting a part with a symmetry tolerance.

Applications of symmetry

The symmetry tolerance is often used to accurately control balance for rotating parts or to insure equal wall thickness. Specify symmetry only when it is necessary because it is time-consuming and expensive to manufacture and inspect. The symmetry control is appropriately used for large, expensive parts that require a small symmetry tolerance to balance mass. If the restrictive symmetry control is not required, a more versatile position tolerance may be used to control a symmetrical relationship. See chapter 8 for a discussion of the application of the position control to tolerance symmetrical features.

Summary

- Concentricity is that condition where the median points of all diametrically opposed points of a surface of revolution are congruent with the axis of a datum feature.

- Concentricity is a location control that has a cylindrical tolerance zone coaxial with the datum axis.

- The concentricity tolerance and datum reference apply only on an RFS basis.

- The aggregate of all median points must lie within a cylindrical tolerance zone whose axis is coincident with the axis of the datum feature.

- The concentricity tolerance is independent of both size and form.
- Differential measurement excludes size, shape, and form while controlling the median points of the feature.
- The concentricity tolerance is often used to accurately control balance for high-speed rotating parts.
- Symmetry is that condition where the median points of all opposed or correspondingly located points of two or more feature surfaces are congruent with the axis or center plane of a datum feature.
- Symmetry is a location control that has a tolerance zone that consists of two parallel planes evenly disposed about the center plane or axis of the datum feature.
- The symmetry tolerance and datum reference apply only at RFS.
- The aggregate of all median points must lie within a tolerance zone defined by two parallel planes equally disposed about the center plane of the datum feature.
- The symmetry tolerance is independent of both size and form.
- The symmetry tolerance is often used to accurately control balance for rotating parts or to insure equal wall thickness.
- Specify symmetry only when it is necessary because it is time-consuming and expensive to manufacture and inspect.

Chapter Review

1. Both concentricity and symmetry controls are reserved for a few _____
   _____.

2. Concentricity and symmetry both employ the same tolerancing _____;
   they just apply to different _____.

3. Concentricity is that condition where the median points of all diametrically opposed points of a surface of revolution are congruent with
   _____
   _____.

4. Concentricity is a _____ control. It has a _____
   tolerance zone that is coaxial with _____.

5. Concentricity tolerance applies only on a _____ basis.
   It must have at least _____that also applies only
   _____.

6. For concentricity, the aggregate of all _____
   must lie within a _____ tolerance zone
   whose axis is coincident with the axis of _____.

7. Concentricity can be inspected, for acceptance only, by placing a
   _____ on the toleranced surface of revolution and rotating
   the part about the _____.

8. To reject parts and to inspect features, such as regular polygons and ellipses, the traditional method of _____ is employed.

9. The concentricity tolerance is often used to accurately control _____ for high-speed rotating parts.

10. Concentricity is time-consuming and expensive to _____ but less expensive to _____ than the runout tolerance.

11. Symmetry is that condition where the _____ of all opposed or correspondingly located points of two or more feature surfaces are_____ with the _____ of a datum feature.

12. Symmetry is a _____control.

13. Symmetry has a tolerance zone that consists of_____ evenly disposed about the _____ of the datum feature.

14. Symmetry tolerance applies only at _____.

15. Symmetry must have at least one _____ that also applies only at _____.

16. The aggregate of all _____ must lie within a tolerance zone defined by _____ equally disposed about the center plane of the _____.

17. The symmetry tolerance is independent of both _____.

18. Differential measurement excludes_____ while controlling the median points of the feature.

19. The symmetry tolerance is often used to accurately control _____ for rotating parts or to insure equal_____.

20. Specify symmetry only when it is necessary because it is _____ _____to manufacture and inspect.

Problems

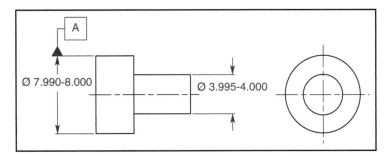

Figure 10-7   Coaxiality of a cylinder: Problem 1.

1. The mass of the high-speed rotating part in Fig. 10-7 must be accurately balanced. The form of the surface is sufficiently controlled by the size tolerance. Specify a coaxiality control for the axis of the 4.000-inch diameter within a tolerance of .001 at RFS to datum A at RFS.

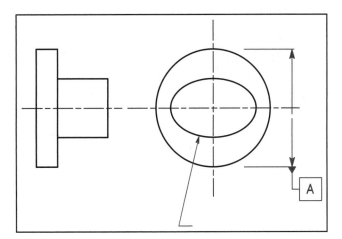

Figure 10-8   Coaxiality of an ellipse: Problem 2.

2. The mass of the ellipse shown in Fig. 10-8 must be accurately balanced. Specify a coaxiality control that will locate the median points of the ellipse within a tolerance of .004 at RFS to datum A at RFS.

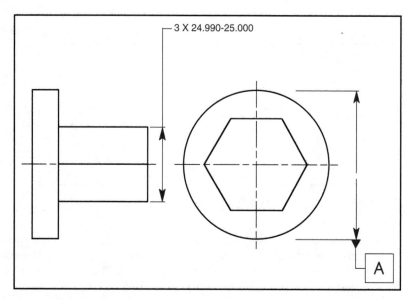

Figure 10-9   Coaxiality of the hexagon: Problem 3.

3. The mass of the hexagon shown in Fig. 10-9 must be accurately balanced. Specify a coaxiality control for the median points of the hexagon within a tolerance of .005 at RFS to datum A at RFS.

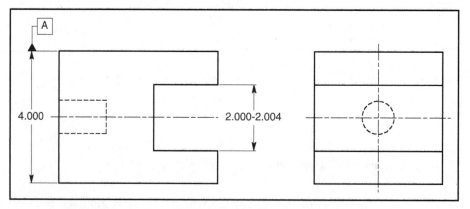

Figure 10-10   Symmetry of the slot: Problem 4.

4. The part in Fig. 10-10 rotates at a high speed, and the mass must be accurately balanced. Specify a geometric tolerance that will centrally locate the slot in this part within a tolerance of .005 at RFS to datum A at RFS.

11

Runout

Runout is a surface control. It controls surfaces constructed around a datum axis and surfaces constructed perpendicular to a datum axis. Runout controls several characteristics of surfaces of revolution, such as coaxiality and circularity, as that surface is rotated about its datum axis.

Chapter Objectives

After completing this chapter, you will be able to

- *Explain* the difference between circular and total runout
- *Specify* runout and partial runout
- *Explain* the application of multiple datum features
- *Explain* the meaning of face and diameter datums
- *Specify* geometric controls to refine datum features
- *Explain* the surface relationship between features controlled with runout
- *Inspect* runout

Definition

Runout is a composite tolerance used to control the functional relationship of one or more features of a part to a datum axis.

Circular Runout

Circular runout applies to every circular element on the surface of a part constructed either around its datum axis or perpendicular to its datum axis, while the part is rotated 360° about that datum axis. Circular runout tolerance

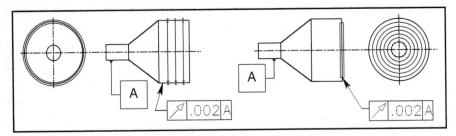

Figure 11-1   Circular runout applied around a datum axis and perpendicular to a datum axis.

applies independently to each circular line element at each measurement position and may easily be applied to cones and curved profiles constructed around a datum axis. Where applied to surfaces constructed around a datum axis, circular runout controls a combination of variations in circularity and coaxiality. Where applied to surfaces at a 90° angle to a datum axis, circular runout controls variations in perpendicularity of circular elements to its datum axis, that is, total runout controls wobble.

Total Runout

Total runout is a compound control that applies to all elements on the surface of a part either around its datum axis or perpendicular to its datum axis, as the part is rotated 360° about that datum axis. Total runout tolerance applies simultaneously to all circular and profile measurement positions. Where applied to surfaces constructed around a datum axis, total runout controls a combination of coaxiality, circularity, straightness, angularity, taper, and profile variations of the surface. Where applied to surfaces at a 90° angle to a datum axis, total runout controls the combination of variations of perpendicularity to the datum axis and flatness, i.e., total runout controls wobble and concavity or convexity.

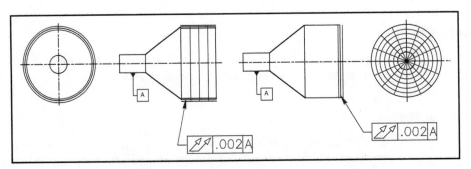

Figure 11-2   Total runout applied around a datum axis and perpendicular to a datum axis.

Specifying Runout and Partial Runout

When specifying runout, the feature control frame is connected to the controlled surface with a leader. In some infrequent instances, the feature control frame may be attached to the extension of a dimension line if the surface to be controlled is small or inaccessible. The feature control frame consists of a runout symbol, the numerical tolerance, and at least one datum. No other symbols are appropriate in the feature control frame. Since runout is a surface control, no material condition applies; consequently, in effect, runout applies at regardless of feature size. Where runout is required for only a portion of a surface, a thick chain line is drawn on one side adjacent to the profile of the surface and dimensioned with a basic dimension as shown in Fig. 11-3.

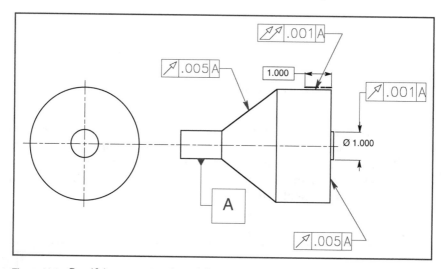

Figure 11-3 Specifying runout and partial runout.

Multiple Datum Features

At least one datum must be specified for a runout control. In many cases, two functional datum diameters are used to support a rotating part such as the one shown in Fig. 11-4. Datum A is no more important than datum B and vice versa. Fig. 11-4 shows datums A and B at a slight angle to the true axis, but taken together, the two datums tend to equalize their axes, producing a single datum axis truer than the axis either one of the datums would have produced independently.

Face and Diameter Datums

The face and diameter datum reference frame specified in Fig. 11-5 is quite a different requirement than the multiple datum feature reference shown in

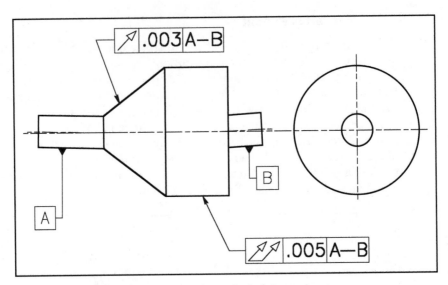

Figure 11-4 Two datum diameters creating a single datum axis.

Fig. 11-4. Datums A and B are specified in two separate compartments in the feature control frame. Therefore, datum A is more important than datum B. That means that the surfaces being controlled must first be perpendicular to datum plane A and then be rotated about datum axis B, even if datum B is not exactly perpendicular to datum A.

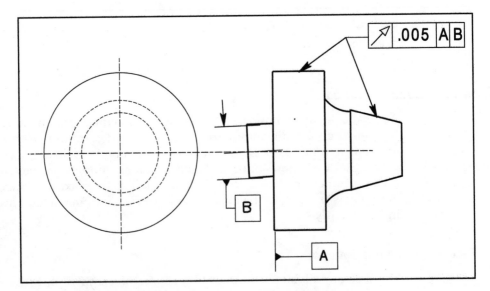

Figure 11-5 Face and diameter datums specified in order of precedence.

Geometric Controls to Refine Datum Features

It may be particularly important for datum features to have a form control refinement. Datums A and B in Fig. 11-6 have a cylindricity refinement of .0005. Design requirements may make it necessary to restrict datum surface variations with respect to straightness, flatness, circularity, cylindricity, and parallelism. It may also be necessary to include a runout control for individual datum features on a multiple datum feature reference, as shown in Fig. 11-6. Datums A and B are independently controlled with a circular runout tolerance to datum A–B. This tolerance is not the same as controlling a feature to itself. In fact, it is expected that datum axis A and datum axis B are coaxial with datum axis A–B, but in the event that datum A or datum B is out of tolerance with respect to datum A–B, the part does not meet design requirements.

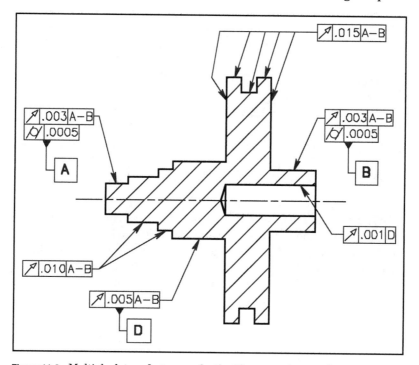

Figure 11-6 Multiple datum features refined with geometric controls.

Surface Relationships Between Features

If two or more surfaces are controlled with a runout tolerance to a common datum reference, the worst-case runout between two surfaces is the sum of the two individual runout tolerances. For example, in Fig. 11-6, the worst-case runout between the largest diameter and datum D diameter is the sum of .015 plus .005 or a total of .020. The two axes may, in fact, be coaxial; however, at

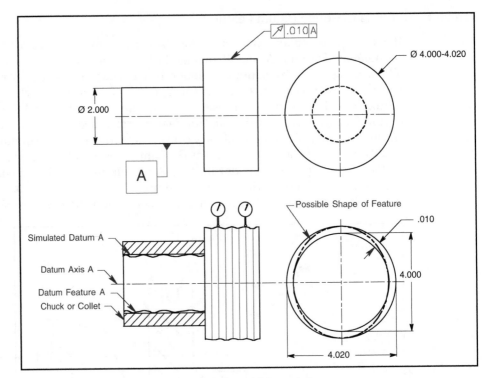

Figure 11-7 Inspecting circular runout relative to a datum axis.

worst case, the surface of the larger diameter could be translated .015 in one direction, and the surface of datum D could also be translated .005 in the other direction producing a difference of .020 between the two surfaces.

If two features have a specific relationship between them, one should be toleranced directly to the other and not through a common datum axis. Figure 11-6 shows a .500 internal diameter controlled directly with a runout tolerance of .001 to datum D rather than controlling each feature to datum A–B. If the Ø .500 hole had been toleranced to datums A–B, the runout tolerance between the hole and datum D would be a total of .006.

In Fig. 11-6, the multiple leaders directed from the .015 circular runout feature control frame to the five surfaces of the part may be specified without affecting the runout tolerance. It makes no difference whether one or several leaders are used with a feature control frame as long as the runout tolerance and datum feature(s) are the same.

Inspecting Runout

When inspecting circular runout, the feature, first, must fall within the specified size limits. It must also satisfy Rule 1, i.e., it may not exceed the boundary of perfect form at maximum material condition. The datum feature is then mounted in a chuck or a collet. With a dial indicator contacting the surface to be inspected, the part is rotated 360° about its simulated datum axis. Several measuring positions are inspected. If the full indicator movement (FIM) does not exceed the specified runout tolerance, the feature is acceptable. Runout tolerance may be larger than the size tolerance. If the runout tolerance is larger than the size tolerance and no other geometric tolerance is applied, the size tolerance controls the form. If the size tolerance is larger than the runout tolerance, circular runout refines circularity as well as controls coaxiality.

The same preliminary checks required for circular runout are also required for total runout. Just as when inspecting circular runout, a dial indicator contacts the surface to be inspected, but the dial indicator is moved along the full length of the feature's profile as the part is rotated 360° about its simulated datum axis. If the FIM does not exceed the specified runout tolerance, the feature is acceptable.

Summary

- Runout is a composite tolerance used to control the functional relationship of one or more features of a part to a datum axis.

- Circular runout applies to every circular element on the surface of a part constructed either around a datum axis or perpendicular to a datum axis, as the part is rotated 360° about its datum axis.

- Total runout is a compound control that applies to all elements on the surface of a part either around a datum axis or perpendicular to a datum axis, as the part is rotated 360° about its datum axis.

- When specifying runout, the feature control frame consists of a runout symbol, the numerical tolerance, and at least one datum. No other symbols are appropriate in the feature control frame.

- Two datum diameters may be used to control a rotating part.

- Face and diameter datums are specified in order of precedence.

- Datum features may have a form control refinement.

- If two or more surfaces are controlled with a runout tolerance to a common datum reference, the worst-case runout between two surfaces is the sum of the two individual runout tolerances.

- If two features have a specific relationship between them, one should be toleranced to the other.

- Several leaders may be used with a single feature control frame as long as the runout tolerance and datum feature(s) are the same.

Chapter Review

1. Circular runout applies to every _____ on the surface of a part constructed either around a datum axis or perpendicular to a datum axis, as the part is rotated _____ about its datum axis.

2. Where circular runout is applied to surfaces constructed around a datum axis, it controls a combination of variations in
   _____.

3. Where circular runout is applied to surfaces at a 90° angle to a datum axis, it controls variations in _____ of circular elements to its datum axis.

4. Total runout is a compound control that applies to all elements on the surface of a part either _____ or _____ , as the part is rotated _____ about its datum axis.

5. Total runout tolerance applies simultaneously to all _____ and _____ measuring position.

6. Total runout applied to surfaces constructed around a datum axis controls a combination of _____ variations of the surface.

7. Total runout applied to surfaces at a 90° angle to a datum axis controls the combination of variations of _____.

8. The runout feature control frame consists of _____
   _____.

9. In many cases, two functional _____ are used to support a rotating part.

10. Where face and diameter datum surfaces are specified, the surface being controlled must first be perpendicular to the _____ datum.

11. Design requirements may make it necessary to restrict datum surface variations with respect to (other geometric controls) _____
    _____.

12. It may be necessary to include a runout control for individual datum features on a _____.

13. If two or more surfaces are controlled with a runout tolerance to a common datum reference, the worst-case runout between two surfaces is _____
    _____.

14. If two features have a specific relationship between them, one should be __
    _____.

15. Multiple leaders directed from a runout feature control frame may be specified without _____.

Problems

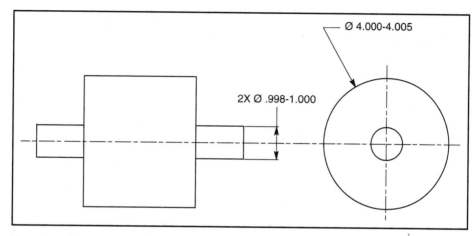

Figure 11-8   Runout control: Problem 1.

1. On the part in Fig. 11-8, control the 4.000-inch diameter with a total runout tolerance of .002 to both the 1.000-inch diameters.

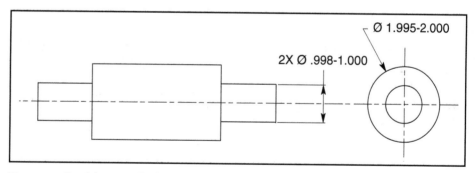

Figure 11-9   Partial runout: Problem 2.

2. On the drawing in Fig. 11-9, specify a circular runout tolerance of .002 controlling the 2.000-inch diameter to both the 1.000-inch diameters. This control is a partial runout tolerance 1.000 inch long starting from the left end of the 2.000-inch diameter. Specify a circular runout of .001 for each of the 1-inch diameters.

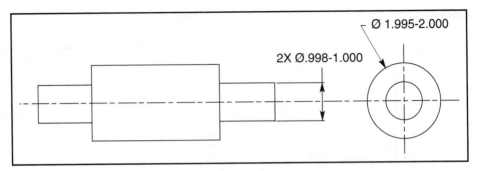

Figure 11-10 Datums toleranced with a cylindricity tolerance: Problem 3.

3. Tolerance the two-inch diameter with a total runout tolerance of .0010 to both the one-inch diameter shafts. Tolerance each one-inch diameter shaft with a cylindricity tolerance of .0005.

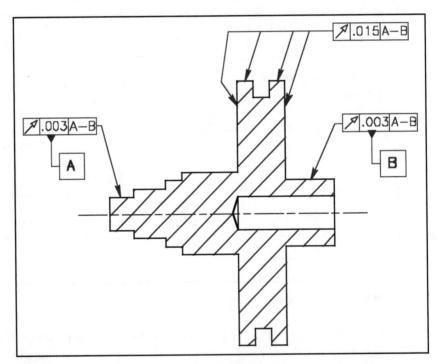

Figure 11-11 Multiple features tolerance with one feature control frame: Problem 4.

4. In Fig. 11-11, which datum, A or B, takes precedence? _____

5. What is the worst possible runout tolerance between the two largest diameters in Fig. 11-11? _____

Profile

Profile is a surface control. It is a powerful and versatile tolerancing tool. It may be used to control just the size and shape of a feature or the size, shape, orientation, and location of an irregular-shaped feature. The profile tolerance controls the orientation and location of features with unusual shapes, very much like the position tolerance controls the orientation and location of holes or pins.

Chapter Objectives

After completing this chapter, you will be able to

- *Specify* a profile tolerance
- *Explain* applications of a profile tolerance zone
- *Properly apply* datums for the profile tolerance
- *Explain* the need for a radius control with a profile
- *Explain* the combination of a profile tolerance with other geometric controls
- *Specify* coplanarity
- *Properly apply* composite profile tolerancing

Definition

A profile is the outline of an object. Specifically, the profile of a line is the outline of an object in a plane as the plane passes through the object. The profile of a surface is the result of projecting the profile of an object on a plane or taking cross sections through the object at various intervals.

Specifying Profile

A profile view or section view of a part is dimensioned with basic dimensions. A true profile may be dimensioned with basic size dimensions, basic coordinate dimensions, basic radii, basic angular dimensions, formulas, or undimensioned drawings. The feature control frame is always directed to the profile surface with a leader. Profile is a surface control; the association of a profile tolerance with an extension or a dimension line is inappropriate. The profile feature control frame contains the profile of a line or of a surface symbol and a tolerance. Since profile controls are surface controls, cylindrical tolerance zones and material conditions do not apply in the tolerance section of profile feature control

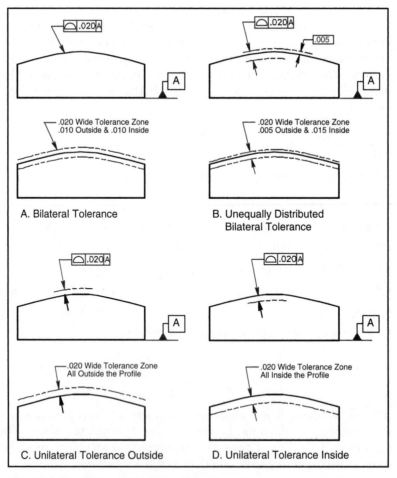

Figure 12-1 Specifying profile of a surface.

frames. The shape of the tolerance zone is the shape of the profile not a cylinder, and material condition modifiers do not apply to surface controls.

When the leader from a profile tolerance points directly to the profile, the tolerance specified in the feature control frame is equally disposed about the true profile. In Fig. 12-1A, the .020 tolerance in the feature control frame is evenly divided, .010 outside and .010 inside the true profile. If the leader from a profile tolerance points directly to a segment of a phantom line extending, outside or inside, parallel to the true profile, as shown in Fig. 12-1C and 12-1D, all the tolerance is outside or inside the true profile. The tolerance may even be specified as an unequal bilateral tolerance by drawing segments of phantom lines inside and outside parallel to the profile and specifying the outside tolerance with a basic dimension, as shown in Fig. 12-1B.

Where a profile tolerance applies all around the profile of a part, the "all around" symbol is specified, as shown in Fig. 12-2A. The "all around" symbol is indicated by a circle around the joint in the leader from the feature control frame to the profile. If the profile is to extend between two points, as shown in Fig. 12-2B, the points are labeled, and a note using the "between" symbol is placed beneath the feature control frame. The profile tolerance applies to the portion of the profile between points X and Z where the leader is pointing. If a part, such as a casting or forging, is to be controlled with a profile tolerance over its entire surface, the note "ALL OVER" is placed beneath the feature control frame, as shown in Fig. 12-2C. When an unusual profile tolerancing requirement occurs, one not covered by the notes and symbols above, a local note clearly stating the extent and application of the profile tolerance must be included.

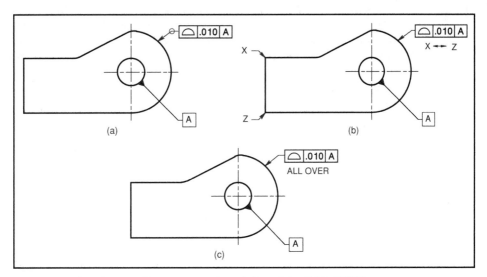

Figure 12-2 The "all around" and "between" symbols and the "ALL OVER" note.

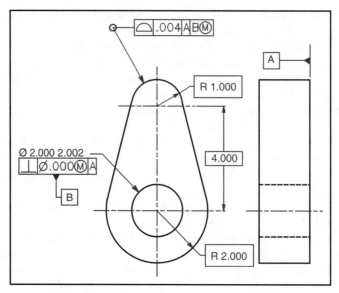

Figure 12-3 The orientation and location of a profile to datum A and to datum B at MMC.

The Application of Datums

Profile tolerances may or may not have datums. The profile of a surface control usually requires a datum to properly orient and locate the surface. The application of datums for the profile control is very similar to the application of datums for the position control. In Fig. 12-3, the profile of a surface is oriented perpendicularly to datum plane A and located to the hole, datum B, at maximum material condition (MMC). Material conditions apply for datums if they are features of size. Datums are generally not used for the profile of a line when only the cross section is being controlled. An example of the application of the profile of a line without a datum would be a profile control specifying a tolerance for a continuous extrusion.

A Radius Refinement with Profile

The profile tolerance around a sharp corner, labeled P in Fig. 12-4, is typically larger than the specified tolerance. Consequently, a sharp corner tolerance will allow a relatively large radius on the part profile. Excessively large radii are shown in Fig. 12-4. If the design requires a smaller radius than the radius allowed by the profile tolerance, a local note such as "R .015 MAX" or "ALL CORNERS R .015 MAX" is directed to the radius with a leader.

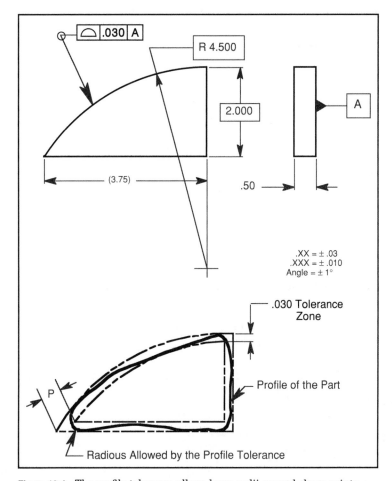

Figure 12-4 The profile tolerance allows large radii around sharp points.

Combing Profile Tolerances with Other Geometric Controls

The profile tolerance may be combined with other geometric tolerances to refine certain aspects of a surface. Examples are given in the drawings below:

In Fig. 12-5, the surface of the profile has a parallelism refinement. Since parallelism only applies to lines and planes, a parallelism control is inappropriate to refine the surface of a profile. But in this example, the parallelism is specified for "EACH ELEMENT" as indicated by the note beneath the feature control frame. While the profile must fall within a .030 tolerance zone, each individual line element in the profile must be parallel to datums A and B within a tolerance zone of .010.

In Fig. 12-6, the surface of the profile has a circular runout refinement. While the profile must fall within a tolerance of .020 about datum axis A–B starting

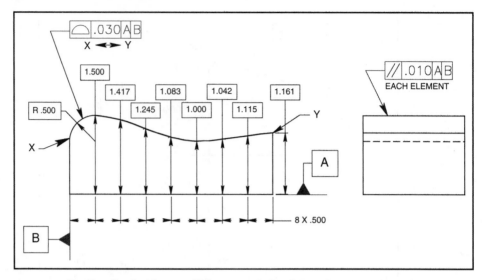

Figure 12-5   Profile refined with a parallelism control of each line element in the surface.

at datum C, each circular element in the profile about the datum axis must also fall within a circular runout of .004 to datum A–B.

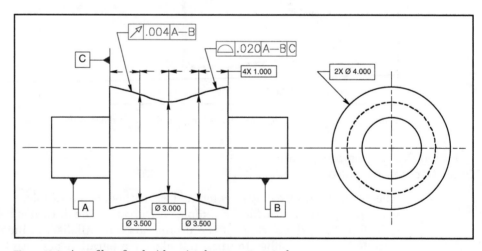

Figure 12-6   A profile refined with a circular runout control.

Coplanarity

Coplanarity is the condition of two or more surfaces having all elements in one plane. Where coplanarity is required, a profile of a surface tolerance is specified. The profile of a surface feature control frame is specified, with a

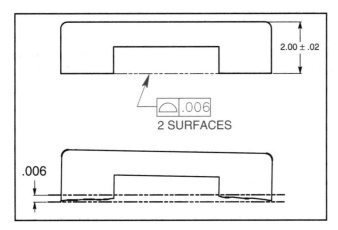

Figure 12-7 Specifying coplanarity with the profile control.

leader, directed to a phantom line connecting the coplanar surfaces. A note indicating the number of coplanar surfaces is placed beneath the feature control frame. As shown in Fig. 12-7, the coplanar surfaces are not necessarily parallel to the opposite (top) surface. However, the size of the feature must be within its specified size tolerance. Coplanarity of two or more surfaces specified with a profile tolerance is similar to flatness of a single surface specified with a flatness tolerance.

When the opposite surface is specified as a datum and the datum is included in the profile feature control frame, as shown in Fig. 12-8, the toleranced surfaces must be coplanar and parallel to the datum surface within the tolerance specified in the feature control frame. If the 2.000-inch plus or minus dimension is specified, as in Fig. 12-8A, the .006 profile tolerance zone must be parallel to datum A and must fall within a .040 size tolerance zone, i.e., the .006 tolerance zone may float up and down within a .040 size tolerance zone but must remain

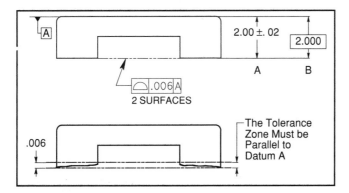

Figure 12-8 Two coplanar surfaces parallel to a datum.

parallel to datum A. Coplanarity of two or more surfaces specified with a profile tolerance including a datum, which identifies a parallel surface, is similar to parallelism of a single surface specified with a parallelism tolerance.

If the basic 2.000-inch dimension is specified, as in Fig. 12-8B, the true profile of the two coplanar surfaces is a basic 2.000 inches from datum A, and the two parallel planes, .006 apart are evenly disposed about the true profile. Where the basic dimension is specified, the total tolerance for the size, parallelism, and coplanarity is .006. Coplanarity of two or more surfaces specified with a profile tolerance, a datum, and a basic dimension is treated like any other profile control.

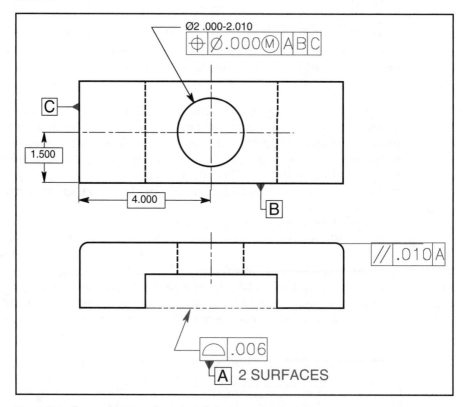

Figure 12-9   Two coplanar surfaces as a datum.

Coplanar surfaces may be used as a datum. If this is the case, it is best to attach the datum feature symbol to the profile feature control frame and include a note specifying the number of coplanar surfaces, as shown in Fig. 12-9.

Profile of a Conical Feature

Conicity may be controlled with a profile tolerance. A conicity tolerance specifies a tolerance zone bounded by two coaxial cones at the specified basic angle

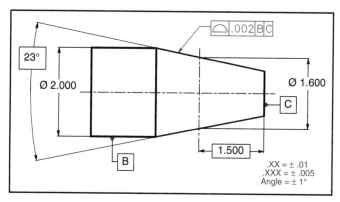

Figure 12-10   Specifying profile of a cone to datum features.

with a radial separation equal to the specified tolerance. The conical feature must fall inside the profile tolerance zone, and it must also satisfy the size tolerance requirements. The size tolerance is specified by identifying a diameter with a basic dimension and tolerancing that diameter with a plus or minus tolerance. If just the form of a cone is to be toleranced, no datums are required. Figure 12-10 shows datums controlling both form and orientation of a cone.

Composite Profile

Composite profile tolerancing is very similar to composite positional tolerancing discussed in chapter 8. A composite profile feature control frame has one profile symbol that applies to the two horizontal segments that follow. The upper segment, called the profile-locating control, governs the location relationship between the datums and the profile. It acts like any other profile control. The lower segment, referred to as the profile refinement control, is a smaller tolerance than the profile-locating control and governs the size, form, and orientation relationship of the profile. The smaller tolerance zone need not fall entirely inside the larger tolerance zone, but any portion of the smaller tolerance zone that lies outside the larger tolerance zone is unusable. The feature profile must fall inside both profile tolerance zones.

For composite profile tolerancing, there is a requirement and a condition:

- Any datums in the lower segment of the feature control frame are *required* to repeat the datums in the upper segment. If only one datum is repeated, it would be the primary datum; if two datums were repeated, they would be the primary and secondary datums, and so on.
- The *condition* of datums in the lower segment of the feature control frame is that they control only orientation.

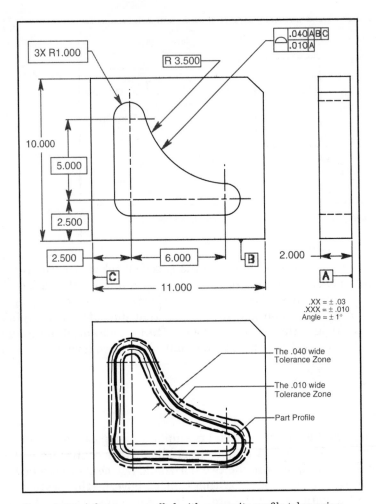

Figure 12-11 A feature controlled with composite profile tolerancing.

The profile in Fig. 12-11 must fall within the .010 tolerance zone governing form and orientation to datum A. The entire profile, however, may float around within the larger tolerance zone of .040 located to datums B and C.

A composite profile may also be used to control orientation to a larger tolerance with a refinement of form to a smaller tolerance in the lower

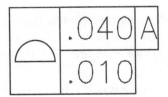

Figure 12-12 Composite profile tolerancing used only to control form and orientation.

segment of the feature control frame shown in Fig. 12-12. The upper segment governs the orientation relationship between the profile and the datum. The lower segment is a smaller tolerance than the profile orienting control and governs the size and form relationship of the profile. The smaller tolerance zone need not fall entirely inside the larger tolerance zone, but any portion of the smaller tolerance zone that lies outside the larger tolerance zone is unusable. The feature profile must fall inside both profile tolerance zones.

A second datum may be repeated in the lower segment of the composite feature control frame, as shown in Fig. 12-13. Both datums in the lower segment of the feature control frame only control orientation. Since datum A in the upper

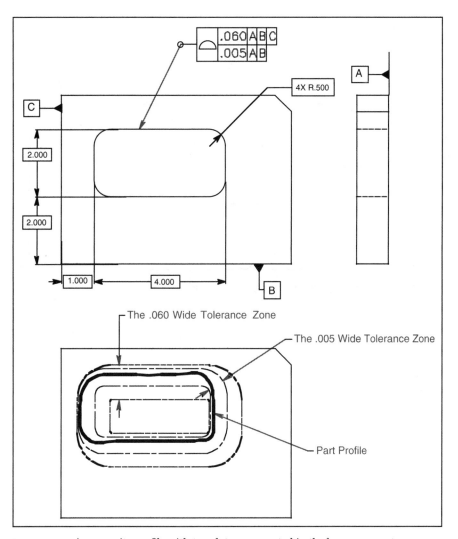

Figure 12-13   A composite profile with two datums repeated in the lower segment.

segment controls only orientation, i.e., perpendicularly to datum A, it is not surprising that datum A in the lower segment is a refinement of perpendicularity to datum A. When datum B is included in the lower segment, the .005-wide tolerance zone must remain parallel to datum B—the smaller tolerance zone is allowed to translate up and down and left and right but may not rotate about an axis perpendicular to datum plane A. The smaller tolerance zone must remain parallel to datum B at all times, as in Fig. 12-13.

The profile in Fig. 12-14 is toleranced with a two single-segment feature control frame. In this example, the lower segment refines the profile just as the lower segment of the composite feature control frame does, but the datums behave differently. The lower segment of a two single-segment feature control

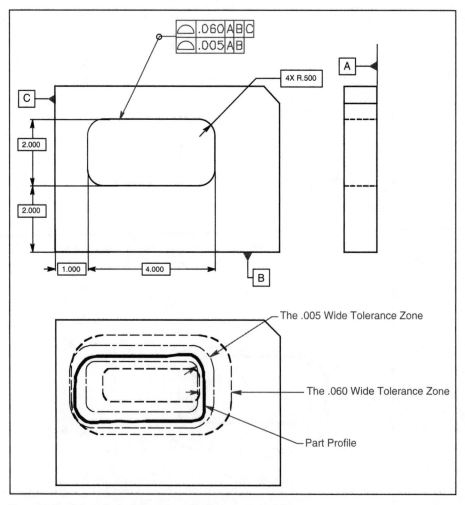

Figure 12-14   A two single-segment profile feature control frame.

frame acts just like any other profile control. If datum C had been included in the lower segment, the upper segment would be meaningless and the entire profile would be controlled to the tighter tolerance of .005. In Fig. 12-14, the lower segment of the two single-segment feature control frame controls profile size, form, orientation, and location to datum B within a .005-wide profile tolerance zone. In other words, the actual profile must fit inside the profile refinement tolerance, be perpendicular to datum A, and be located a basic 2.000 inches from datum B within a tolerance of .005. The upper segment, the profile locating control, allows the .005-wide profile refinement tolerance zone to translate back and forth within a profile tolerance of .060, i.e., the refinement tolerance zone may translate left and right but may not translate up and down or rotate in any direction.

Inspection

Inspecting a surface that has been controlled with a profile tolerance can be accomplished in a number of ways. The most common methods of inspecting a profile are listed below:

- A gage made to the extreme size and shape of the profile can be used.
- A thickness gage can be used to measure variations between a template, made to the true size and shape of the profile, and the actual surface.
- An open setup with a dial indicator can be used to inspect some profiles.
- An optical comparator is designed to inspect profiled surfaces. An optical comparator projects a magnified projected outline on to a screen. The projected outline is then compared to a profile template.
- Some coordinate measuring machines are designed to inspect profile.

Summary

- **A** profile is the outline of an object.
- The true profile may be dimensioned with basic size dimensions, basic coordinate dimensions, basic radii, basic angular dimensions, formulas, or undimensioned drawings.
- A profile is a surface control.
- When the leader from a profile feature control frame points directly to the profile, the tolerance specified in the feature control frame is equally disposed about the true profile.
- The "all around" symbol is indicated by a circle around the joint in the leader.
- If the profile is to extend between two points, the points are labeled, and a note using the "between" symbol is placed beneath the feature control frame.

- Profile tolerances may or may not have datums.
- The profile tolerance may be combined with other geometric tolerances to refine certain aspects of a surface.
- Where coplanarity is required, a profile of a surface tolerance is specified.
- Composite profile tolerancing is very similar to composite positional tolerancing.
- Datums in the lower segment of a composite feature control frame control only orientation.
- A profile may be toleranced with a two single-segment feature control frame.

Chapter Review

1. Profile of a line is the _____
   of an object in a plane as the plane passes through the object.

2. Profile of a surface is the result of _____
   _____ .
   or taking cross sections through the object at various intervals.

3. The true profile may be dimensioned with what kind of dimensions?_____
   _____
   _____

4. The feature control frame is always directed to the profile surface with a
   _____ .

5. What symbols do not apply in the tolerance section of profile feature control frames? _____

6. When the leader from a profile tolerance points directly to the profile, the tolerance specified in the feature control frame is _____
   _____ .

7. If the leader from a profile tolerance points directly to a segment of a phantom line extending, outside or inside, parallel to the profile, then ____
   _____ .

8. Where a profile tolerance applies all around the profile of a part, the _____
   _____ is specified.

9. Draw the "all around" symbol. _____

10. If the profile is to extend between two points, the points are _____
    and a note using the _____ is placed beneath the feature control frame.

11. Draw the "between" symbol. _____

12. If a part is to be controlled with a profile tolerance over its entire surface, the note_____ is placed _____.

13. Profile tolerances _____ have datums.

14. The profile of a surface control usually requires a datum to properly _____ _____.

15. Datums are generally_____for the profile of a line when only_____is being controlled.

16. If the design requires a smaller radius than the radius allowed by the profile tolerance, a note such as _____ is directed to the radius with a _____.

17. The profile tolerance may be combined with other _____ to _____ certain aspects of a surface.

18. Coplanarity is the condition of _____ surfaces having all _____.

19. Coplanarity is toleranced with the profile of a surface feature control frame, connected with a _____, to a _____ connecting the surfaces.

20. Where specifying coplanarity, a note indicating _____ is placed beneath the _____.

21. Where coplanar surfaces are used as a datum, it is best to attach the datum feature symbol to _____and _____.

22. Conicity may be controlled with a _____.

23. Composite profile tolerancing is very similar to _____ _____.

24. The upper segment of a composite profile feature control frame is called the _____ and it governs the _____.

25. The lower segment, referred to as the _____, is a smaller tolerance than the profile locating control and governs _____ _____of the profile.

26. The feature profile must fall inside _____.

27. For composite profile tolerancing, there is a requirement and a condition: _____ _____. _____ _____.

28. A second datum may be repeated in the lower segment of the composite feature control frame that also controls _____ .

29. The lower segment of a two single-segment feature control frame acts just like _____ .

30. The upper segment of a two single-segment feature control frame allows the smaller tolerance zone to _____ relative to the datum not
repeated in the lower segment within the larger tolerance.

Problems

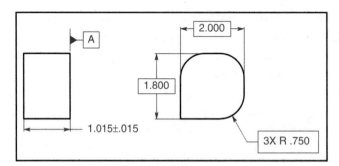

Figure 12-15   Profile of a surface: Problem 1.

1. Specify the profile of a surface tolerance of .020, perpendicular to datum A, and all around the part in Fig. 12-15.

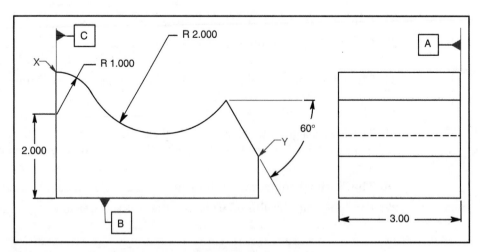

Figure 12-16   Profile of a surface between two points: Problem 2.

2. For the curved surface and angle in Fig. 12-16, specify the profile of a surface tolerance of .030, located to datums A, B, and C, between points X and Y.

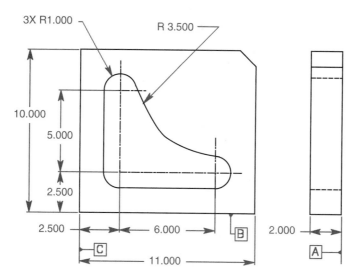

Figure 12-17 Locating profile of a surface: Problem 3.

3. Control the entire surface of the center cavity to the datums indicated within a tolerance of .015 outside the true profile. (Outside the profile is external to the true profile line. Inside the profile is within the profile loop.)

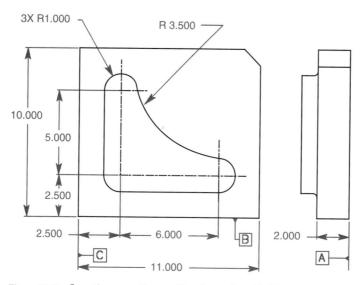

Figure 12-18 Locating a mating profile of a surface: Problem 4.

4. Control the entire surface of the punch to the datums indicated within a tolerance of .015 inside the true profile. (Outside the profile is external to the true profile line. Inside the profile is within the profile loop.)

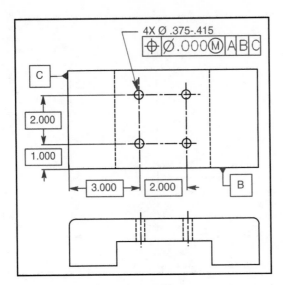

Figure 12-19 Coplanarity: Problem 5.

5. The primary datum is the two lower coplanar surfaces. Specify the primary datum to be coplanar within .004.

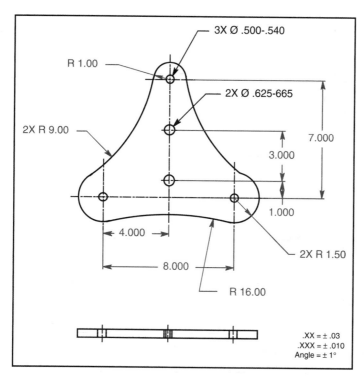

3X Ø .500-.540

R 1.00

2X Ø .625-665

2X R 9.00

7.000

3.000

1.000

4.000

8.000

2X R 1.50

R 16.00

.XX = ± .03
.XXX = ± .010
Angle = ± 1°

Figure 12-20 Profile controlled to size feature datums: Problem 6.

6. Specify controls locating the hole-patterns to one another and perpendicular to the back of the part. Specify a control locating the profile to the hole patterns and perpendicular to the back of the part within a tolerance of .060. The holes are for 1/2-inch and 5/8-inch bolts.

7. Specify a profile tolerance for the center cutout that will control the size and orientation to datum A within .010, and locate it to the datums indicated within .060. Complete the drawing in Fig. 12-21.

8. Draw a profile tolerance that will satisfy the requirements for problem 7, and *orient* the cutout parallel to datum B within .010.

9. Draw a profile tolerance that will satisfy the requirements for problem 7, and *locate* the cutout to datum B within .010.

10. Specify the bottom of the lower surface of the sheet metal part in Fig. 12-22 coplanar within .020. Tolerance holes with geometric tolerancing. The smallest size tolerance for each hole is the virtual condition for the mating part. Specify the profile of the top surface of the part within .040.

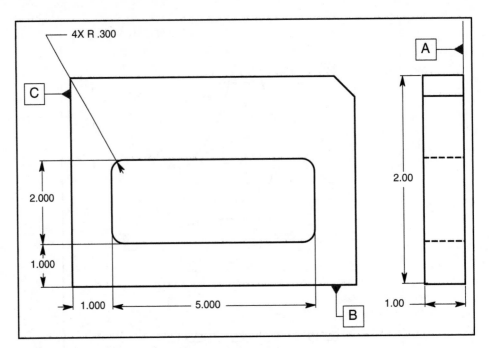

Figure 12-21 A composite profile—Problems 7 through 9.

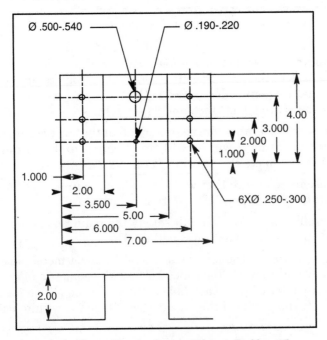

Figure 12-22 The profile of a sheet metal part: Problem 10.

Graphic Analysis

Graphic analysis, sometimes referred to as paper gaging, is a technique that effectively translates coordinate measurements into positional tolerance geometry that can easily be analyzed. It provides the benefit of functional gaging without the time and expense required to design and manufacture a close-tolerance, hardened-metal functional gage.

Chapter Objectives

After completing this chapter, you will be able to

- *Identify* the advantages of graphic analysis
- *Explain* the accuracy of graphic analysis
- *Perform* inspection analysis of a composite geometric tolerance
- *Perform* inspection analysis of a pattern of features controlled to a datum feature of size

Advantages of Graphic Analysis

The graphic analysis approach to gaging has many advantages compared to gaging with traditional functional gages. A partial list of advantages would include the following:

- Provides functional acceptance: Most hardware is designed to provide interchangeability of parts. As machined features depart from their maximum material condition (MMC) size, location tolerance of the features can be increased while maintaining functional interchangeability. The graphic analysis technique provides an evaluation of these added functional tolerances in the acceptance process. It also shows how an unacceptable part can be reworked.

- Reduces cost and time: The high cost and long lead time required for the design and manufacture of a functional gage can be eliminated in favor of graphic analysis. Inspectors can conduct an immediate, inexpensive functional inspection at their workstations.

- Eliminates gage tolerance and wear allowance: Functional gage design allows 10 percent of the tolerance assigned to the part to be used for gage tolerance. Often, an additional wear allowance of up to 5 percent will be designed into the functional gage. This could allow up to 15 percent of the part's tolerance to be assigned to the functional gage. The graphic analysis technique does not require any portion of the product tolerance to be assigned to the verification process. Graphic analysis does not require a wear allowance since there is no wear.

- Allows functional verification of MMC, RFS, and LMC: Functional gages are primarily designed to verify parts toleranced with the MMC modifier. In most instances, it is not practical to design functional gages to verify parts specified at RFS or LMC. With the graphic analysis technique, features specified with any one of these material condition modifiers can be verified with equal ease.

- Allows verification of a tolerance zone of any shape: Virtually a tolerance zone of any shape (round, square, rectangular, etc.) can easily be constructed with graphic analysis methods. On the other hand, hardened-steel functional gaging elements of nonconventional configurations are difficult and expensive to produce.

- Provides a visual record for the material review board: Material review board meetings are postmortems that examine rejected parts. Decisions on the disposition of nonconforming parts are usually influenced by what the most senior engineer thinks or the notions of the most vocal member present rather than the engineering information available. On the other hand, graphic analysis can provide a visual record of the part data and the extent that it is out of compliance.

- Minimizes storage required: Inventory and storage of functional gages can be a problem. Functional gages can corrode if they are not properly stored. Graphic analysis graphs and overlays can easily be stored in drawing files or drawers.

The Accuracy of Graphic Analysis

The overall accuracy of graphic analysis is affected by such factors as the accuracy of the graph and overlay gage, the accuracy of the inspection data, the completeness of the inspection process, and the ability of the drawing to provide common drawing interpretations.

An error equal to the difference in the coefficient of thermal expansion of the materials used to generate the data graph and the tolerance zone overlay gage may be encountered if the same materials are not used for both sheets. Paper also expands with the increase of humidity and its use should be avoided. Mylar is a relatively stable material; when used for both the data graph and the tolerance zone overlay gage, any expansion or contraction error will be nullified.

Layout of the data graph and tolerance zone overlay gage will allow a small percentage of error in the positioning of lines. This error is minimized by the scaling factor selected for the data graph.

Analysis of a Composite Geometric Tolerance

A pattern of features controlled with composite tolerancing can be inspected with a set of functional gages. Each segment of the feature control frame represents a gage. To inspect the pattern of holes in Fig. 13-1, the pattern-locating control, the upper segment of the feature control frame, consists of three mutually perpendicular planes, datums A, B, and C, and four virtual condition pins .242 in diameter. The feature-relating control, the lower segment of the feature control frame, consists of only one plane, datum A, and four virtual condition pins .250 in diameter. These two gages are required to inspect this

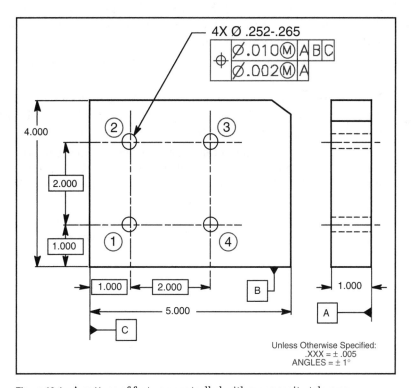

Figure 13-1 A pattern of features controlled with a composite tolerance.

TABLE 13-1   Inspection Data Derived from a Part Made from Specifications in the Drawing in Fig. 13-1

Feature number	Feature location from datum C X-axis	Feature location from datum B Y-axis	Feature size	Departure from MMC (bonus)	Datum-to-pattern tolerance zone size	Feature-to-feature tolerance zone size
1	.997	1.003	Ø.256	.004	Ø.014	Ø.006
2	1.004	3.004	Ø.258	.006	Ø.016	Ø.008
3	3.006	2.998	Ø.260	.008	Ø.018	Ø.010
4	3.002	.998	Ø.254	.002	Ø.012	Ø.004

pattern. If gages are not available, graphic analysis can be used. The procedure for inspecting composite tolerancing with graphic analysis is presented below.

The following is the sequence of steps for generating a **data graph** for the graphic analysis of a composite tolerance:

1. Collect the inspection data shown in Table 13-1.

2. On a piece of graph paper, select an appropriate scale, and draw the specified datums. This sheet is called the data graph. The drawing, the upper segment of the composite feature control frame, and the inspection data dictate the configuration of the data graph.

3. From the drawing, determine the true position of each feature, and draw the centerlines on the data graph.

4. Since tolerances are in the magnitude of thousandths of an inch, a second scale, called the deviation scale, is established. Typically, one square on the graph paper equals .001 of an inch on the deviation scale.

5. Draw the appropriate diameter tolerance zone around each true position by using the deviation scale. For the drawing in Fig. 13-1, each tolerance zone is a circle with a diameter of .010 plus its bonus tolerance. The datum-to-pattern tolerance zone diameters are listed in Table 13-1.

6. Draw the actual location of each feature axis on the data graph. If the location of any of the feature axes falls outside the feature's respective circular tolerance zone, the datum-to-pattern relationship is out of tolerance and the

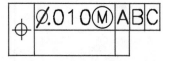

Figure 13-2 The upper segment of the composite feature control frame in Fig. 13-1.

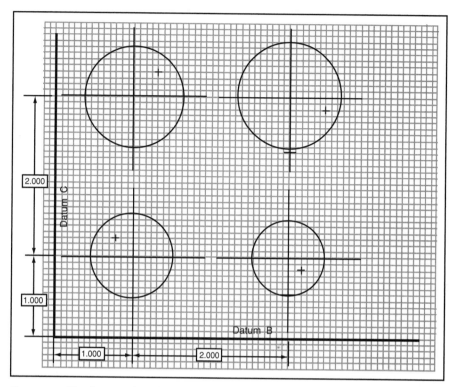

Figure 13-3 The data graph with tolerance zones and feature axes for the data in Table 13-1.

part is rejected. If all of the axes fall inside their respective tolerance zones, the datum-to-pattern relationship is in tolerance, but the pattern must be further evaluated to satisfy the feature-to-feature relationships.

The following is the sequence of steps for generating a tolerance zone overlay gage for the graphic analysis evaluation of a composite tolerance:

1. Place a piece of tracing paper over the data graph. Trace the true position axes on the tracing paper. This sheet is called the tolerance zone overlay gage. The drawing, the lower segment of the feature control frame, and

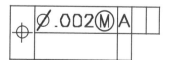

Figure 13-4 The lower segment of the composite feature control frame in Fig. 13-1.

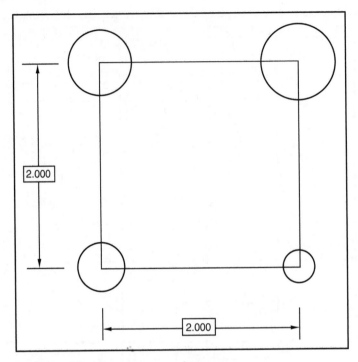

2.000

2.000

Figure 13-5 The tolerance zone overlay gage.

the inspection data dictate the configuration of the tolerance zone overlay gage.

2. Draw the appropriate feature-to-feature positional tolerance zones around each true position axis on the tracing paper. Each tolerance zone is a circle with a diameter of .002 plus its bonus tolerance. The feature-to-feature tolerance zone diameters are listed in Table 13-1.

3. If the tracing paper can be adjusted to include all actual feature axes within the tolerance zones on it, the feature-to-feature relationships are in tolerance. If each axis simultaneously falls inside both of its respective tolerance zones, the pattern is acceptable.

When the tolerance zone overlay gage is placed over the data graph in Fig. 13-6, the axes of holes 1 through 3 can easily be placed inside their respective tolerance zones. The axis of the fourth hole, however, will not fit inside the fourth tolerance zone. Therefore, the pattern is not acceptable. It is easy to see on the data graph that this hole can be reworked. Simply enlarging the fourth hole by about .004 will make the pattern acceptable.

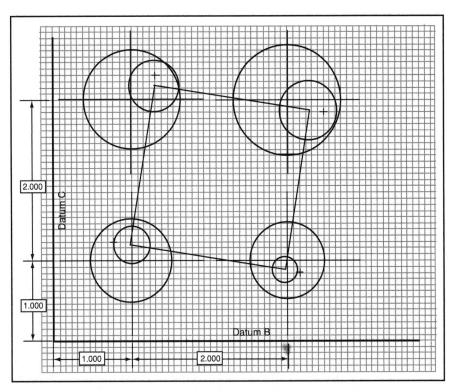

Figure 13-6 The tolerance zone overlay gage is placed on top of the data graph.

Analysis of a Pattern of Features Controlled to a Datum Feature of Size

A pattern of features controlled to a datum feature of size specified at MMC is a very complicated geometry that can easily be inspected with graphic analysis.

The following is the sequence of steps for generating a data graph for the graphic analysis evaluation of a pattern of features controlled to a datum feature of size:

1. Collect the inspection data shown in Table 13-2.

2. On the data graph, select an appropriate scale, and draw the specified datums. The drawing, the feature control frame controlling the hole pattern, and the inspection data dictate the configuration of the data graph.

3. From the drawing, determine the true position of the datum feature and the true position of each feature in the pattern. Draw their centerlines on the data graph.

4. Establish a deviation scale. Typically one square on the graph paper equals .001 of an inch on the deviation scale.

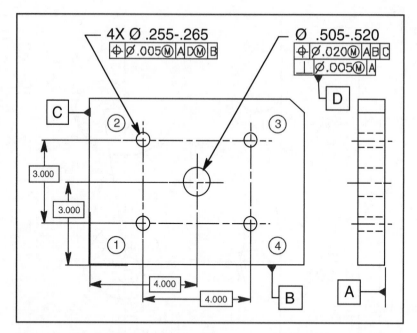

Figure 13-7 The drawing of a pattern of features controlled to a datum feature of size.

5. Draw the appropriate diameter tolerance zone around each true position using the deviation scale. For the drawing in Fig. 13-7, each tolerance zone is a circle with a diameter of .005 plus its bonus tolerance. The total geometric tolerance diameters are listed in Table 13-2.

6. Draw the actual location of each feature on the data graph. If each feature axis falls inside its respective tolerance zone, the part is in tolerance. If one or more feature axes fall outside their respective tolerance zones, the part may still be acceptable if there is enough shift tolerance to shift all the axes into their respective tolerance zones.

TABLE 13-2 Inspection Data Derived from a Part Made from Specifications in the Drawing in Fig. 13-7

Feature number	Feature location from datum D X-axis	Feature location from datum D Y-axis	Actual feature size	Departure from MMC (bonus)	Total geometric tolerance
1	−1.997	−1.498	Ø.258	.003	Ø.008
2	−1.998	1.503	Ø.260	.005	Ø.010
3	2.005	1.504	Ø.260	.005	Ø.010
4	2.006	− 1.503	Ø.256	.001	Ø.006
Datum			Ø.510	Shift Tolerance = .010	

$\oplus$ $\varnothing$.005Ⓜ A DⓂ B

Figure 13-8 The feature control frame controlling the four-hole pattern in Fig. 13-7.

If any of the feature axes falls outside its respective tolerance zone, further analysis is required. The following is the sequence of steps for generating an overlay gage for the graphic analysis evaluation of a pattern of features controlled to a datum feature of size:

1. Place a piece of tracing paper over the data graph. This sheet is called the overlay gage.

2. Trace the actual location of each feature axis on to the overlay gage.

3. Trace the true position axis of datum feature D on to the overlay gage.

4. Trace datum plane B on to the overlay gage.

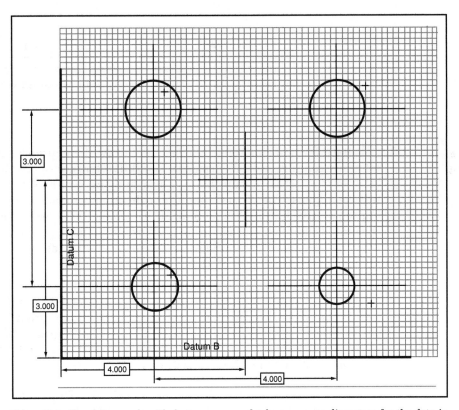

Figure 13-9 The data graph with feature axes and tolerance zone diameters for the data in Table 13-2.

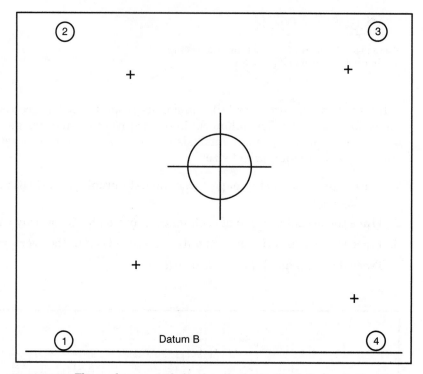

Figure 13-10 The overlay gage includes the actual axis of each feature in the pattern, the shift tolerance zone, and the clocking datum.

5. Calculate the shift tolerance allowed, and draw the appropriate cylindrical tolerance zone around datum axis D. The shift tolerance equals the difference between the actual datum feature size and the size at which the datum feature applies. The virtual condition rule applies to datum D in Fig. 13-7. Consequently, datum D is .505 at MMC minus .005 (geometric tolerance) that equals .500 (virtual condition). According to the inspection data, datum hole D is produced at a diameter of .510. The shift tolerance equals .510 minus .500 or a diameter of .010.

6. If the tracing paper can be adjusted to include all the feature axes on the overlay gage within its' shift tolerance zones on the data graph and datum axis D contained within its shift tolerance zone while orienting datum B on the overlay gage parallel to datum B on the data graph, the pattern of features is in tolerance. The graphic analysis in Fig. 13-11 indicates that the four-hole pattern of features is acceptable.

Graphic analysis is a powerful graphic tool for analyzing part configuration. This graphic tool is easy to use, accurate, and repeatable. It should be in every

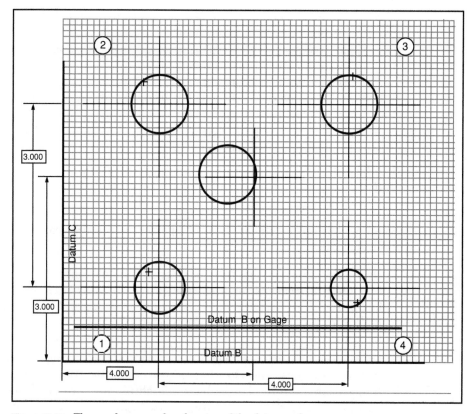

Figure 13-11   The overlay gage placed on top of the data graph.

inspector's bag of tricks. Graphic analysis is also a powerful analytical tool engineers can use to better understand how tolerances on drawings will behave.

Summary

- The advantages of graphic analysis:
  - Provides functional acceptance
  - Reduces time and cost
  - Eliminates gage tolerance and wear allowance
  - Allows functional verification of RFS, LMC, as well as MMC
  - Allows verification of a tolerance zone of any shape
  - Provides a visual record for the material review board
  - Minimizes storage required for gages
- The accuracy of graphic analysis:
  The accuracy of graphic analysis is affected by such factors as the accuracy of the graphs and overlay gage, the accuracy of the inspection data, the

completeness of the inspection process, and the ability of the drawing to provide common drawing interpretations.

- Sequence of steps for the analysis of composite geometric tolerance:

1. Draw the datums, the true positions, the datum-to-pattern tolerance zones, and the actual feature locations on the data graph.

2. On a piece of tracing paper placed over the data graph, trace the true positions, and construct the feature-to-feature tolerance zones. This sheet is called the tolerance *zone* overlay gage.

3. Adjust the tolerance zone overlay gage to fit over the actual feature locations. If each actual feature location falls inside both of its respective tolerance zones, the pattern of features is in tolerance.

Sequence of steps for the analysis of a pattern of features controlled to a datum feature of size:

1. Draw the datums, the true positions, the tolerance zones, and the actual feature locations on the data graph. If the actual feature locations fall inside the tolerance zones, the part is good, and no further analysis is required. Otherwise, continue to step two to utilize the available shift tolerance.

2. On a piece of tracing paper placed over the data graph, trace the actual feature locations, the clocking datum, and the true position of the datum feature of size. Then, draw the shift tolerance zone about the true position of the datum feature of size. This sheet is called the overlay gage.

3. Adjust the overlay gage to fit over the actual feature locations while keeping the shift tolerance zone over the axis on the data gage and the clocking datums aligned. If each actual feature location falls inside both of its respective tolerance zones, the pattern of features is in tolerance.

Chapter Review

1. List the advantages of graphic analysis.

   _____

   _____

   _____

   _____

   _____

   _____

   _____

   _____

2. List the factors that affect the accuracy of graphic analysis.

_____

_____

_____

_____

⊕	⌀.010Ⓜ	A	B	C
	⌀.002Ⓜ	A		

Figure 13-12 Refer to the feature control frame for questions 3 through 7.

3. A piece of graph paper with datums, true positions, tolerance zones, and actual feature locations drawn on it is called a _____ .

4. A piece of tracing paper with datums, true positions, tolerance zones, and actual feature locations traced or drawn is called a _____ .

5. The upper segment of the composite feature control frame, the drawing, and the inspection data dictates the configuration of the _____ .

6. The lower segment of the feature control frame, the drawing, and the inspection data dictate the configuration of the _____ .

7. If the tracing paper can be adjusted to include all feature axes within the _____ on the tracing paper, the feature-to-feature relationships are in tolerance.

⊕	⌀.005Ⓜ	A	DⓂ	B

Figure 13-13 Refer to the feature control frame for questions 8 through 11.

8. To inspect a datum feature of size, the feature control frame, the drawing, and the inspection data dictate the configuration of the _____ .

9. Draw the actual location of each feature on the data graph. If each feature axis falls inside its respective tolerance zone, the part is _____ .

10. If any of the feature axes falls outside its respective tolerance zone, _____ .

_____

11. If the tracing paper can be adjusted to include all the feature axes within the tolerance zones on the data graph and the datum axis contained within its tolerance zone while keeping the pattern parallel to datum B, the pattern of features is _____ .

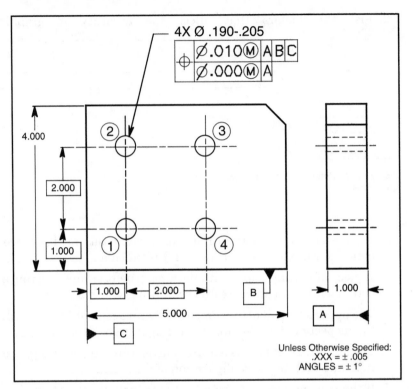

Figure 13-14   A pattern of features controlled with a composite tolerance: Problem 1.

TABLE 13-3   Inspection Data for Graphic Analysis of Problem 1

Feature number	Feature location from datum C X-axis	Feature location from datum B Y-axis	Feature size	Departure from MMC (bonus)	Datum-to-pattern tolerance zone size	Feature-to-feature tolerance zone size
1	1.002	1.003	Ø.200			
2	1.005	3.006	Ø.198			
3	3.005	3.002	Ø.198			
4	3.003	.998	Ø.196			

Problems

1. A part was made from the drawing in Fig. 13-14; the inspection data was tabulated in Table 13-3. Perform a graphic analysis of the part. Is the pattern within tolerance? _____

   If it is not in tolerance, can it be reworked? If so, how?_____

   _____

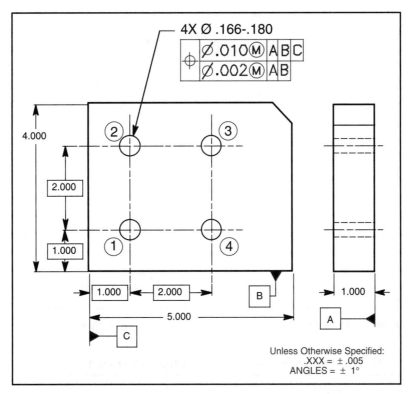

Figure 13-15 A pattern of features controlled with a composite tolerance: Problem 2.

TABLE 13-4 Inspection Data for Graphic Analysis of Problem 2

Feature number	Feature location from datum C X-axis	Feature location from datum B Y-axis	Feature size	Departure from MMC (bonus)	Datum-to-pattern tolerance zone size	Feature-to-feature tolerance zone size
1	1.004	.998	Ø.174			
2	.995	3.004	Ø.174			
3	3.000	3.006	Ø.172			
4	3.006	1.002	Ø.176			

2. A part was made from the drawing in Fig. 13-15; the inspection data was tabulated in Table 13-4. Perform a graphic analysis of the part. Is the pattern within tolerance? _____

   If it is not in tolerance, can it be reworked? If so, how? _____

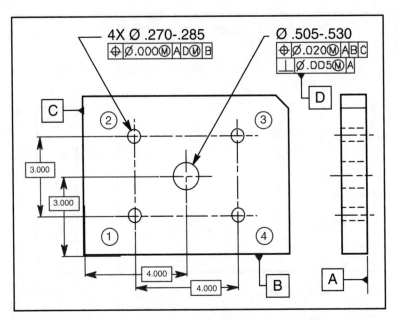

Figure 13-16 A pattern of features controlled to a size feature: Problem 3.

TABLE 13-5   Inspection Data for Graphic Analysis of Problem 3

Feature number	Feature location from datum D X-axis	Feature location from datum D Y-axis	Actual feature size	Departure from MMC (bonus)	Total geometric tolerance
1	−1.992	−1.493	Ø.278		
2	−1.993	1.509	Ø.280		
3	2.010	1.504	Ø.280		
4	2.010	−1.490	Ø.282		
Datum			Ø.520	Shift Tolerance =	

3. A part was made from the drawing in Fig. 13-16; the inspection data was tabulated in Table 13-5. Perform a graphic analysis of the part. Is the pattern within tolerance? _____

If it is not in tolerance, can it be reworked? If so, how? _____

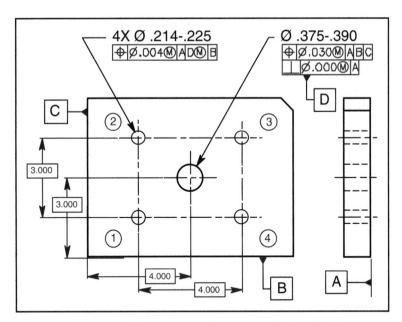

Figure 13-17 A pattern of features controlled to a size feature: Problem 4.

TABLE 13-6   Inspection Data for Graphic Analysis of Problem 4

Feature number	Feature location from datum D X-axis	Feature location from datum D Y-axis	Actual feature size	Departure from MMC (bonus)	Total geometric tolerance
1	−1.995	−1.495	Ø.224		
2	−1.996	1.503	Ø.218		
3	2.005	1.497	Ø.220		
4	1.997	−1.506	Ø.222		
Datum			Ø.380	Shift Tolerance =	

4. A part was made from the drawing in Fig. 13-17; the inspection data was tabulated in Table 13-6.

Perform a graphic analysis of the part. Is the pattern within tolerance? ____

If it is not in tolerance, can it be reworked? If so, how? _____

A Strategy for Tolerancing Parts

When tolerancing a part, the designer must determine the attributes of each feature or pattern of features and the relationships of these features with one another. In other words, what are the size, the size tolerance, the location dimensions, and the location and orientation tolerances of each feature? At what material conditions do these size features apply? Which are the most appropriate datum features? All of these questions must be answered in order to properly tolerance a part. Some designers believe that parts designed with a solid modeling CAD program do not require tolerancing. A note in the Dimensioning and Tolerancing standard indicates caution when designing parts with solid modeling. The standard reads: "CAUTION: If CAD/CAM database models are used and they do not include tolerances, then tolerances must be expressed outside of the database to reflect design requirements." One way or another, each feature must be toleranced.

Chapter Objectives

After completing this chapter, you will be able to

- *Tolerance* size features located to plane surface features
- *Tolerance* size features located to size features
- *Tolerance* a pattern of features located to a second pattern of features

Size Features Located to Plane Surface Features

The first step in tolerancing a size feature, such as the hole in Fig. 14-1, is to specify the size and size tolerance of the feature. The size and the size tolerance may be determined by using one of the fastener formulas, a standard fit table, or the manufacturer's specifications. The second step is locating and orienting the size feature. The location tolerance comes from the size tolerance. If a

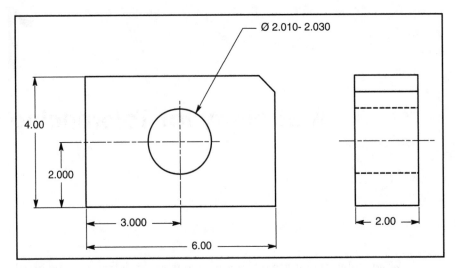

Figure 14-1   A size feature located to plane surfaces on an untoleranced drawing.

Ø 2.000-inch mating feature must fit through the hole in Fig. 14-4, the location tolerance can be as large as the difference between the Ø 2.010 hole and the Ø 2.000-inch mating feature or a positional tolerance of Ø .010. A positional tolerance for locating and orienting a feature of size is always specified with a material condition modifier. The maximum material condition modifier (circle M) has been specified for the hole in Fig. 14-4. The MMC modifier is typically specified for features in static assemblies. The RFS modifier is typically used for high-speed, dynamic assemblies. The LMC modifier is used where a specific minimum edge distance must be maintained. The size tolerance not only controls the feature's size but also controls the feature's form (Rule #1). According to the drawing in Fig. 14-1, the size tolerance for the Ø 2.000-inch hole can be as large as .020. The machinist can make the hole diameter anywhere between 2.010 and 2.030. However, if the machinist actually produces the hole at Ø 2.020, according to Rule #1, the form tolerance for the hole is .010, that is, 2.020 minus 2.010. The hole must be straight and round within .010. The hole size can be produced even larger, up to Ø 2.030, in which case the form tolerance is even larger. If the straightness or circularity tolerance, automatically implied by Rule #1, does not satisfy the design requirements, an appropriate form tolerance must be specified.

The next step in tolerancing a size feature is to identify the location datums. The hole in Fig. 14-1 is dimensioned up from the bottom edge and over from the left edge. Consequently, the bottom and left edges are implied location datums. When geometric dimensioning and tolerancing is applied, these datums must be specified. If the designer has decided that the bottom edge is more important to the part design than the left edge, the datum letter for the bottom edge, datum B, will precede the datum letter for the left edge, datum C, in the feature control frame.

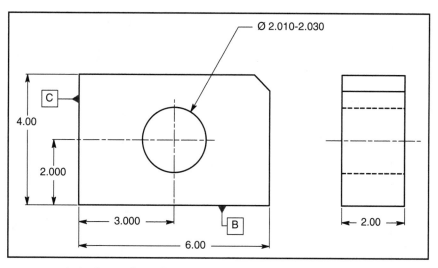

Figure 14-2 A size feature located to specified datums.

Datums B and C not only control location; they also control orientation. If the hole in Fig. 14-2 is controlled with the feature control frame in Fig. 14-3, the hole is to be parallel to datum surfaces B and C within the tolerance specified in the feature control frame.

The primary datum controls orientation with a minimum of three points of contact with the datum reference frame. The only orientation relationship between the hole and datums B and C is parallelism. Parallelism can be controlled with the primary datum but in only one direction. The secondary datum must make contact with the datum reference frame with a minimum of two points of contact; only two points of contact are required to control parallelism in one direction. If the feature control frame in Fig. 14-3 is specified to control the hole in Fig. 14-2, the cylindrical tolerance zone is located from and parallel to datum surfaces B and C, establishing both location and orientation for the feature.

Typically, the front or back surface, or both, is a mating surface, and the hole is required to be perpendicular to one of these surfaces. If that is the case, a third datum feature symbol is attached to the more important of the two surfaces, front or back. In Fig. 14-4, the back surface has been identified as datum A. Since datum A is specified as the primary datum in the feature control frame and the primary datum controls orientation, the cylindrical tolerance zone of the hole is perpendicular to datum A. When applying geometric dimensioning and tolerancing, all datums are identified, basic location dimensions are included, and a feature control frame is specified.

Figure 14-3 A position tolerance locating and orienting the feature to datums B and C.

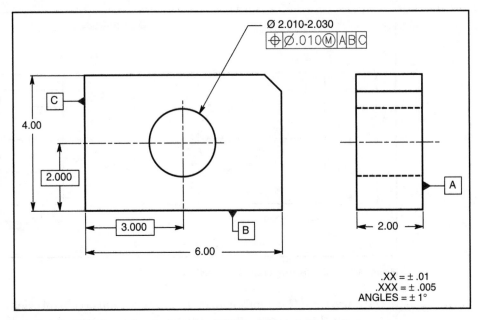

Figure 14-4 A hole located and oriented at MMC to datums A, B, and C.

The primary datum is the most important datum and is independent of all other features—it is not related to any other feature. Other features are controlled to the primary datum. The primary datum is often a large flat surface that mates with another part, but many parts do not have flat surfaces. A large, functional, cylindrical surface may be selected as a primary datum. Other surfaces are also selected as primary datums even if they require datum targets to support them. In the final analysis, the key points in selecting a primary datum are:

- Select a functional surface,
- Select a mating surface,
- Select a sufficiently large, accessible surface that will provide repeatable positioning stability in a datum reference frame while processing and ultimately in assembly

The only appropriate geometric tolerance for a primary datum is a form tolerance. All other tolerances control features to other features. On complicated parts, it is possible to have a primary datum oriented or located to some other feature(s) involving another datum reference frame. However, in most cases, it is best to have only one datum reference frame.

Rule #1 controls the flatness of datum A in Fig. 14-5 if no other control is specified. The size tolerance, a title block tolerance of ±.01, a total tolerance of .020, controls the form. If Rule #1 does not sufficiently control the flatness, a flatness

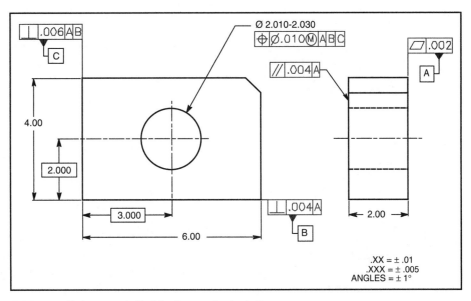

Figure 14-5 Datums controlled for form and orientation.

tolerance must be specified. If the side opposite datum A must be parallel within a smaller tolerance than the tolerance allowed by Rule #1, a parallelism control must be specified, as shown in Fig. 14-5. If required, a parallelism control can also be specified for the sides opposite datums B and C.

In Fig. 14-5, datum B is specified as the secondary datum. The secondary datum is the more important of the two location datums. It may be more important because it is larger than datum C or because it is a mating surface. When producing or inspecting the hole, datum feature B must contact the datum reference frame with a minimum of two points of contact. The perpendicularly of datums B and C to datum A and to each other is controlled by the ±1° angularity tolerance in the title block if not otherwise toleranced. However, as shown in Fig. 14-5, datum B is controlled to datum A with a perpendicularity tolerance of .004. Datum C is specified as the tertiary (third) datum, and it is the least important datum. When producing or inspecting the hole, datum feature C must contact the datum reference frame with a minimum of one point of contact. The orientation of datum C may be controlled to both datums A and B. For the Ø 2.000-inch hole in Fig. 14-5, datum A is the reference for orientation (perpendicularity), and datums B and C are the references for location.

If the Ø .010 tolerance specified for the hole location is also acceptable for orientation, the feature control frame specified in Fig. 14-5 is adequate. If an orientation refinement of the hole is required, a smaller perpendicularity tolerance, such as the one in Fig. 14-6, is specified.

If the hole is actually produced at Ø 2.020, there is a .010 bonus tolerance that applies to both the location and orientation tolerances. Consequently, the *total*

Figure 14-6 A location tolerance with a perpendicularity refinement.

positional tolerance is Ø .020, i.e., a combination of location and orientation may not exceed a cylindrical tolerance of .020. The *total* orientation tolerance may not exceed Ø .010.

The same tolerancing techniques specified for the single hole in the drawings above also apply to a pattern of holes shown in Fig. 14-7. The hole pattern is located with basic dimensions to datum reference frame A, B, and C. The features in the pattern are located to one another with basic dimensions. The note "4X Ø .510–.540" indicates that all four holes have the same size and size tolerance. The geometric tolerance specified beneath the note indicating the hole diameters also applies to all four holes. Each hole in the pattern is positioned and oriented to the datum reference frame within a cylindrical tolerance zone .010 in diameter at MMC.

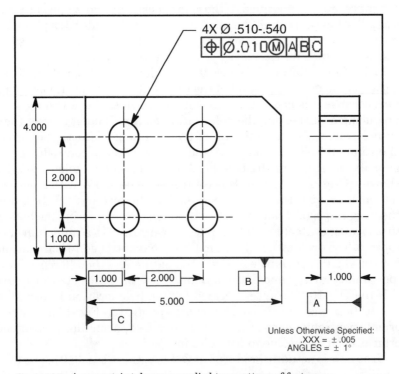

Figure 14-7 A geometric tolerance applied to a pattern of features.

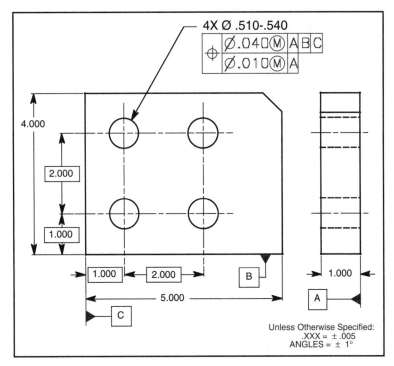

Figure 14-8   A composite positional tolerance applied to a pattern of features.

Composite geometric tolerancing is employed when the tolerance between the datums and the pattern is not as critical as the tolerance between the features within the pattern. This tolerancing technique is often used to reduce the cost of a part. The position symbol applies to both the upper and lower segments of a composite feature control frame. The upper segment controls the *pattern* in the same way that a single feature control frame controls a pattern. The lower segment refines the *feature-to-feature* location relationship; the primary function of the position tolerance is location.

The pattern in Fig. 14-8 is located with basic dimensions to datum reference frame A, B, and C within four cylindrical tolerance zones .040 in diameter at MMC. The relationship between the features located to one another with basic dimensions as well as the perpendicularity to datum A is controlled by four cylindrical tolerance zones .010 in diameter at MMC. The axis of each feature must fall completely inside both of its respective tolerance zones.

Size Features Located to Size Features

Another common geometry with industrial applications is a pattern of holes located to a size feature such as an inside or an outside diameter.

In Fig.14-9, an eight-hole pattern is placed on a basic Ø 2.500 bolt circle, with a basic 45° angle between each feature. The pattern is perpendicular to datum A

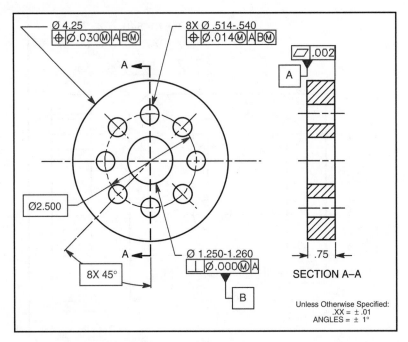

Figure 14-9 A pattern of holes located to a datum feature of size.

and located to datum B, i.e., the center of the bolt circle is positioned on the axis of the center hole, datum B. If the back of this part is to mate with another part and these holes are clearance holes used to bolt the parts together, the holes should be perpendicular to the mating surface. Consequently, it is appropriate to make the back surface of this part the primary datum. It is often necessary to refine the flatness of mating surfaces. Datum surface A has been controlled with a flatness tolerance of .002, which is relatively easy to achieve on a 5 or 6-inch diameter surface.

If the hole pattern were located to the outside diameter, a datum feature symbol would have been attached to the circumference of the part. Many designers indiscriminately pick the outside diameter as a datum feature instead of selecting datum features that are critical to fit and function. Since the inside diameter is the critical feature, the datum feature symbol is attached to the feature control frame identifying the inside diameter as datum B.

Frequently, the secondary datum is controlled perpendicular to the primary datum, but controlling the orientation is even more important if the secondary datum is a size feature like a hole. Not only can the hole be out of perpendicularly, but the mating shaft can also be out of coaxiality with the hole. Datum B has been assigned a zero perpendicularity tolerance at MMC. Since all of the tolerance comes from the bonus, the virtual condition and the MMC are the same diameter. If the machinist produces datum B at a diameter of 1.255, the hole must be perpendicular to datum A within a cylindrical tolerance zone of .005.

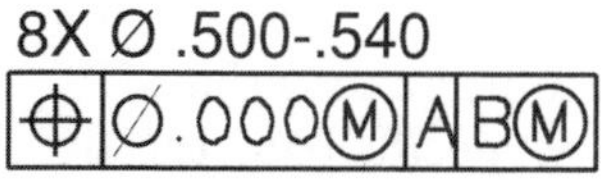

8X Ø .500-.540

⊕ | Ø.000Ⓜ | A | BⓂ

Figure 14-10 A zero positional tolerance for a pattern of holes.

Some designers use position instead of perpendicularity to control the orientation of the secondary datum to the primary datum. This is inappropriate since the secondary datum is not being located to anything. Designers communicate best when they use the proper control for the job.

Finally, the clearance holes are toleranced. If half-inch fasteners are used with a positional tolerance of .014, the MMC hole size is .514. The fastener formula is as follows:

$$\text{Fastener @ MMC} + \text{Geometric tolerance @ MMC} = \text{Hole diameter @ MMC}$$

$$.500 + .014 = .514$$

Positional tolerance for *clearance holes* is essentially arbitrary. The positional tolerance could be .010, .005, or even .000. If zero positional tolerance at MMC were specified, the diameter of the hole at MMC would be .500, as shown in Fig. 14-10.

The hole size at LMC was selected with drill sizes in mind. A Ø 17/32 (.531) drill might produce a hole that is a few thousands oversize resulting in a diameter of perhaps .536. A Ø.536 hole falls within the size tolerance of .514–.540 with a bonus of .022 and a total tolerance of .036. Had the location tolerance been specified at zero positional tolerance at MMC, the Ø.536 hole would still have fallen within the size tolerance of .500–.540 with a bonus of .036. The total tolerance would have been the same, .036. For clearance holes, the positional tolerance is arbitrary.

Since clearance holes imply a static assembly, the MMC modifier (circle M) placed after the tolerance is appropriate. There is no reason the fastener must be centered in the clearance hole; consequently, an RFS material condition is not required. The MMC modifier will allow all of the available tolerance; it will accept more parts and reduce costs.

The primary datum, datum A, is the orientation datum. Datum A, in the positional feature control frame of the hole pattern, specifies that the cylindrical tolerance zone of each hole must be perpendicular to datum plane A. Datum plane A is the plane that contacts a minimum of three high points of the back surface of the part. The secondary datum, datum B, is the locating datum. Datum B is the axis of the Ø 1.250 hole. The center of the bolt circle is located on this datum B axis. Datum B is specified with an MMC modifier (circle M) in the feature control frame. As the size of datum B departs from Ø 1.250 toward Ø 1.260, the pattern gains shift tolerance in the exact amount of such departure. In this particular situation, the virtual condition applies (see the virtual condition rule), but the virtual condition and the MMC are the same since zero perpendicularity at MMC was specified for the datum B hole. If datum hole B is produced at Ø 1.255, there is a cylindrical tolerance zone .005 in diameter about the axis of datum B within which the axis of the bolt circle may shift. In other words, the pattern, **as a whole,** may shift in any direction within a cylindrical tolerance zone .005 in diameter. Shift tolerance may be determined with graphic analysis techniques discussed in chapter 13.

One of the most common drawing errors is the failure to control coaxiality. The feature control frame beneath the Ø 4.25 size dimension controls the coaxiality of the outside diameter to the inside diameter. Coaxiality may be toleranced in a variety of ways, but it must be controlled to avoid incomplete drawing requirements. Many designers omit this control, claiming that it is "over-kill," but sooner or later, they will buy a batch of parts that will not assemble because the features are out of coaxiality.

Some designs require patterns of features to be clocked to a third datum feature. That means, where the pattern is not allowed to rotate about a center axis, a third datum feature is used to prevent rotation.

The pattern of holes in Fig. 14-11 is toleranced in the same way the hole pattern in Fig. 14-9 is toleranced except it has been clocked to datum C. The

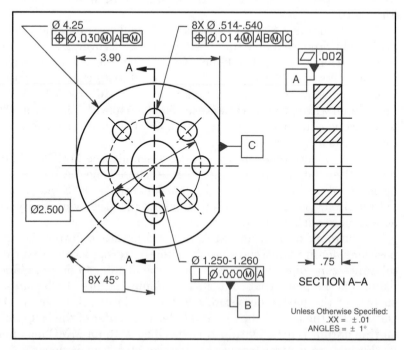

Figure 14-11   A pattern of holes located to a datum feature of size and clocked to a flat surface.

flat on the outside diameter has been designated as datum C and specified as the tertiary datum in the feature control frame, preventing clocking of the hole pattern about datum B. Some designers want to control datum surface C perpendicular to the horizontal axis passing through the hole pattern, but datum C is THE DATUM. The horizontal axis passing through the hole pattern must be perpendicular to datum C, not the other way around.

Many parts have a clocking datum that is a size feature such as a hole or keyseat. The pattern of holes in Fig. 14-12 is toleranced in the same way as the hole pattern in Fig. 14-9 except that it has been clocked to datum C, which in this case is a size feature. Datum C is a .500-inch keyseat with its own geometric tolerance. The keyseat is perpendicular to the back surface of the part and located to the 1.250 diameter hole within a tolerance zone of two parallel planes .000 apart at MMC. The keyseat gains tolerance as the feature departs from .500 toward .510 wide. The center plane of the keyseat must fall between the two parallel planes.

The hole pattern is clocked to datum C at MMC. The virtual condition rule applies, but since the control is a zero positional tolerance, both the MMC and the virtual condition are the same—.500. If the keyseat is actually produced at a width of .505, the hole pattern has a shift tolerance of .005 with respect to datum C. That means that the entire pattern can shift up and down and

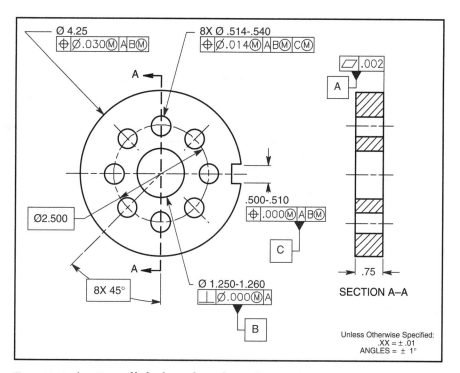

Figure 14-12 A pattern of holes located to a datum feature of size and clocked to a keyseat.

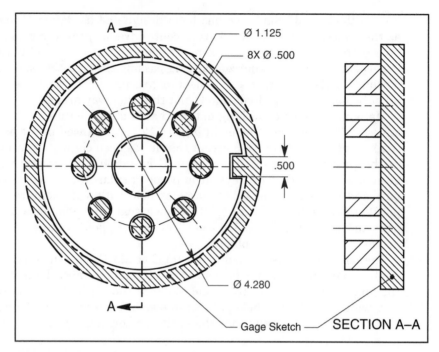

Figure 14-13 A gage sketched about the part in Fig. 14-12 illustrates a shift tolerance.

can clock within the .005 shift tolerance zone. This is assuming that there is sufficient shift tolerance available from datum B. If there is little or no shift tolerance from datum B, datum C will only allow a clocking shift around datum B.

Tolerances on parts like the one in Fig. 14-12 are complicated and sometimes difficult to visualize. It is helpful to draw the gage that would inspect the part. It is not difficult; on a print, just make a sketch around the part. This sketch is sometimes called a "cartoon gage." The sketch illustrates how the part must first sit flat on its back surface, datum A. It is easy to see how the part can shift about the 1.125 center diameter, datum B, and the .500 key, datum C. Finally, the outside diameter of the part must be sufficiently coaxial to fit inside the 4.280 diameter. Visualization of shift tolerances can be greatly enhanced with the use of a gage sketch.

A Pattern of Features Located to a Second Pattern of Features

Individual features and patterns of features may be toleranced to patterns of features and individual size features. There are several ways of specifying datums to control the two patterns of features in Fig. 14-14.

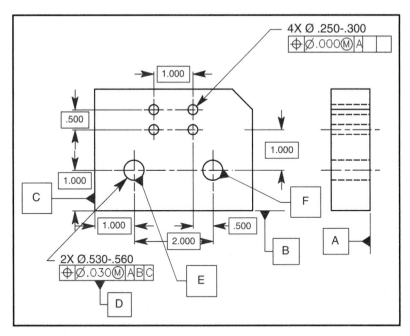

Figure 14-14 A pattern of holes located to a second pattern of holes.

In Fig. 14-14, the .500-inch hole pattern is positioned to plane surface datums. The cylindrical tolerance zones of the holes are perpendicular to datum A, located up from datum B and over from datum C.

Now that the two-hole pattern is toleranced, what is the best way to tolerance the four-hole pattern? The simplest and most straightforward way of tolerancing the four-hole pattern is to control it to datums A, B, and C—Fig. 14-15. Where possible, it is best to use only one datum reference frame. In this example, the patterns are controlled to each other through datum reference frame A, B, and C. If both hole patterns are toleranced to the same datums, in the same order of precedence, and at the same material conditions, the patterns are to be considered one composite pattern of features. Since one pattern has a cylindrical tolerance of .030 at MMC and the other has a cylindrical tolerance of .000 at MMC, the two patterns will be located to each other within a cylindrical tolerance of .030 at MMC. If the tolerance between patterns must be smaller then Ø .030, it can be reduced.

⊕	Ø .000 Ⓜ	A	B	C

Figure 14-15 A feature control frame controlling the four-hole pattern to datums A, B, and C.

⊕ | ⌀.000 Ⓜ | A | D Ⓜ

Figure 14-16   A feature control frame locating the four-hole pattern to datum D at MMC.

If a large location tolerance for the two-hole pattern from datums A, B, and C and a small tolerance between the two-hole and four-hole patterns is desirable, one of the patterns must be the locating datum. In Fig. 14-14, the two-hole pattern—both .500-inch holes—is identified with a datum feature symbol as one datum, datum D.

If the four-hole pattern is controlled with the feature control frame in Fig. 14-16, the four-hole pattern is to be perpendicular to datum A and located to datum D at MMC within the tolerance specified, i.e., both holes in the two-hole pattern act as one datum controlling the location and clocking of the four-hole pattern. The part is shown in Fig. 14-17 on a gage designed to inspect the four-hole pattern perpendicular to datum A and located to the two-hole pattern, datum D.

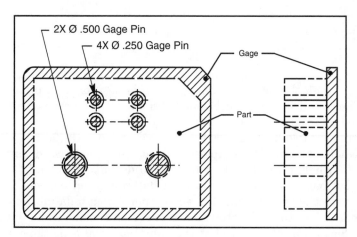

Figure 14-17   A gage locating the four-hole pattern to the two-hole pattern, datum D.

The feature control frame in Fig. 14-18 is equivalent to the feature control frame in Fig. 14-16. If the four-hole pattern on the drawing is controlled with the feature control frame in Fig. 14-18, it is to be perpendicular to datum A and located to the datum E at circle M–F at circle M within the tolerance specified. Datums E and F are of equal value. Datum E at MMC–datum F at MMC in the feature control frame for the four-hole pattern will produce the same gage as datum D at MMC. The gage in Fig. 14-17 will inspect the four-hole pattern controlled with either of the feature control frames in Fig. 14-16 or Fig. 14-18.

⊕ | ⌀.000 Ⓜ A | E Ⓜ –F Ⓜ

Figure 14-18 A feature control frame locating the four-hole pattern to datum E at MMC dash datum F at MMC.

The feature control frame in Fig. 14-19 is similar to the feature control frame in Fig. 14-18, except that datum E, the secondary datum, is more important than datum F, the tertiary datum. As a result, datum E is the locating datum, and datum F is the clocking datum, i.e., the function of datum F is only to prevent the part from rotating about datum E.

⊕ | ⌀.000 Ⓜ A | E Ⓜ | F Ⓜ

Figure 14-19 A feature control frame locating the four-hole pattern to datum E at MMC and datum F at MMC.

The gage in Fig. 14-20 is designed to inspect the four-hole pattern when it is located to datum E at MMC and clocked to datum F at MMC. Notice that the datum F pin on the gage is diamond-shaped. The diamond-shaped pin allows contact only at the top and bottom edges in order to limit clocking about datum pin E. Both are virtual condition pins, but the diamond-shaped pin is only the virtual condition of the hole along its vertical axis. No other parts of the pin may touch the inside of the hole.

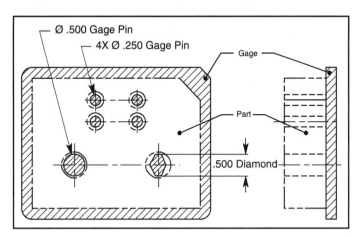

Figure 14-20 A gage locating the four-hole pattern to datum E at MMC and clocking to datum F at MMC.

Of the tolerancing techniques discussed above, the simplest is the plane surface datum reference frame, datums A, B, and C. If the two holes are the locating datum, use the datum D technique. If only one hole is the datum, specify that hole as the locating datum, and specify another feature, such as datum F or datum B, as the clocking datum.

Summary

- A designer must determine the attributes of each feature and the relationship between the features.

- First, specify the size and size tolerance of a feature.

- Determine whether the size tolerance controls the feature's form (Rule #1) or whether a form tolerance is required.

- Identify the datums and the order in which they appear in the feature control frame.

- The primary datum is the most important datum and is not controlled to any other feature. If Rule #1 does not sufficiently control the form of the primary datum, a form tolerance must be specified.

- Perpendicularity controls of the secondary and tertiary datums must be specified if the title block angularity tolerance is not adequate.

- The same tolerancing techniques specified for a single hole also apply to a pattern of holes.

- Composite geometric tolerancing is employed when the tolerance between the datums and the pattern is not as critical as the tolerance between the features within the pattern.

- Another common geometry with industrial applications is a pattern of holes located to a size feature such as an inside or an outside diameter. Typically, the pattern is perpendicular to a flat surface, datum A, and located to a size feature, datum B.

- Frequently, the secondary datum is controlled to the primary datum with a perpendicularity tolerance.

- One of the fastener formulas is used to calculate the positional tolerance of clearance holes.

- For clearance holes, the positional tolerance at MMC is arbitrary. A zero positional tolerance at MMC is as good as, if not better than, specifying a tolerance in the feature control frame.

- If the center of a bolt circle is located on the axis of a datum feature of size and the datum feature is specified with an MMC modifier, the pattern of features gains shift tolerance as the center datum feature of size departs from MMC toward LMC.

- A pattern of features may be clocked to a tertiary datum, such as a flat or a keyseat, to prevent rotation about the secondary datum.

- The simplest and most straightforward way of tolerancing multiple patterns of features is to use a plain surface datum reference frame, if possible.

- A second choice is to specify one pattern as the datum.

- A third choice is to choose one feature in the pattern as the locating datum and another feature as a clocking datum.

Chapter Review

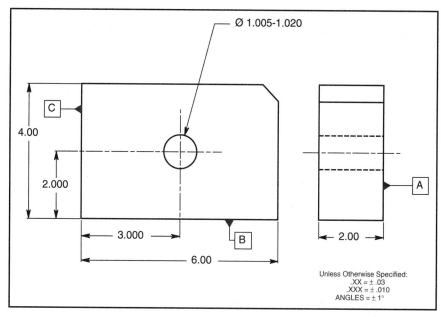

Figure 14-21 A hole located and oriented to datums A, B, and C for questions 1 through 5.

1. What category of geometric tolerances applies to the primary datum in a drawing like the drawing in Fig. 14-21? _____

2. What geometric tolerance applies to the primary datum in the drawing in Fig. 14-21? _____

3. The primary datum controls _____ of the feature being controlled.

4. Assume the feature control frame for the hole in Fig. 14-21 happens to be:

    What relationship would the Ø 1.005 hole have to datums B and C?

    _____
    _____
    _____

5. Assume the feature control frame for the hole in Fig. 14-21 happens to be:

    What relationship would the Ø 1.005 hole have to datums A, B, and C?

    _____
    _____
    _____

6. Complete the feature control frame below so that it refines orientation to .000 at MMC.

7. Draw a feature control frame to control a pattern of holes within Ø .125 at MMC to its datums A, B, and C. Refine the tolerance of the feature-to-feature relationship to Ø .000 at MMC.

8. What is the orientation tolerance for the pattern of holes in the answer that you specified for question number 7? _____

9. Keeping in mind that the primary datum controls orientation, explain how you would select a primary datum on a part. _____
_____
_____
_____

10. How would you determine which datum should be secondary and which tertiary?
_____
_____
_____

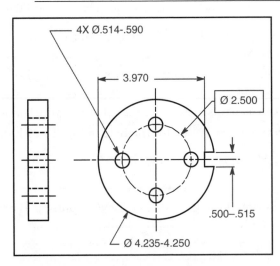

Figure 14-22 Pattern of features for questions 11 through 17.

11. Select a primary datum, and specify a form control for it.

12. Select a secondary datum, and specify an orientation control for it. The virtual condition of the mating inside diameter is Ø 4.250.

13. Tolerance the keyseat for a .500-inch key.

14. Tolerance the .500-inch clearance holes for .500-inch floating fasteners.

15. Are there other ways this part can be toleranced?

16. If the outside diameter is actually produced at 4.240, how much shift tolerance is available?

17. If the outside diameter is actually produced at 4.240 and the keyseat at .505, how much can this part actually shift? Sketch a gage about the part.

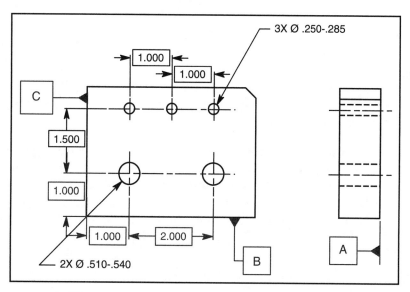

Figure 14-23   Two patterns of features for questions 18 through 21.

18. Locate the two-hole pattern to the surface datums with a positional tolerance of Ø .085 at MMC. Locate the two holes to each other, and orient them to datum A within a tolerance of Ø .010 at MMC.

19. Locate the three-hole pattern to the two-hole pattern within a Ø .000 positional tolerance.

20. The two-hole pattern is specified as a datum at MMC; at what size do the two holes apply?_____

21. What is the total possible shift tolerance allowed for the three-hole pattern?

Problems

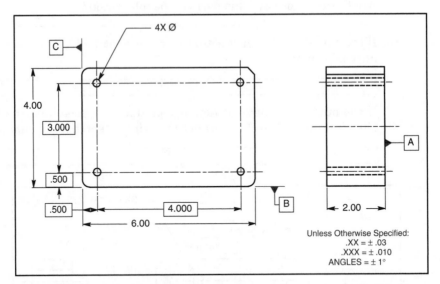

Figure 14-24   Tolerancing: Problem 1.

1. Dimension and tolerance the four-hole pattern for #10 cap screws as fixed fasteners. Allow maximum tolerance for the clearance holes and 60 percent of the total tolerance for the threaded holes in the mating part.
   - How flat is datum surface A?_____
   - How perpendicular are datums B and C to datum A and to each other?_____

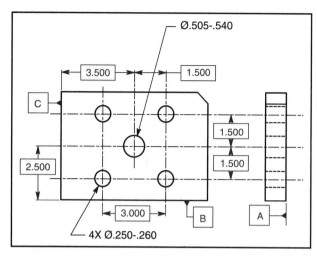

Figure 14-25   Tolerancing: Problem 2.

2. Tolerance the center hole to the outside edges with a tolerance of .060. Refine the orientation of the .005 hole to the back of the part within .005. Control the four-hole pattern to the center hole. The four-hole pattern mates with a part having four pins with a virtual condition of .250. Give each feature all of the tolerance possible.

   ■ At what size does the center hole apply for the purposes of positioning the four-hole pattern?_____

   ■ If the center hole is produced at a diameter of .535, how much shift of the four-hole pattern is possible?_____

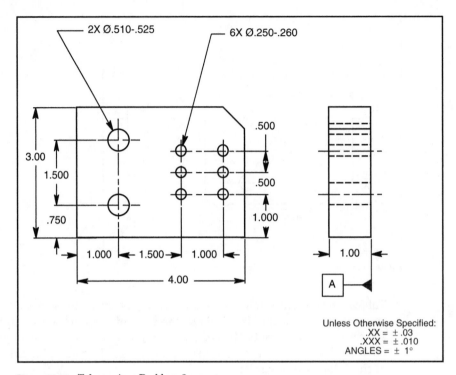

Figure 14-26   Tolerancing: Problem 3.

3. The location of the hole patterns to the outside edges is not critical; a tolerance of .060 at MMC is adequate. The location between the two .500-inch holes and their orientation to datum A must be within .010 at MMC. Control the six-hole pattern to the two-hole pattern within .000 at MMC. The mating part has virtual condition pins of Ø .500 and Ø.250.

- At what size does the two-hole pattern apply for the purposes of positioning the six-hole pattern?_____
- If the two large holes are produced at a diameter of .540, how much shift of the four-hole pattern is possible?_____

Appendix

TABLE A-1 Oversize Diameters in Drilling*

Drill diameter, inch	Amount oversize, inch		
	Average max.	Mean	Average min.
1/16	.0020	.0015	.0010
1/8	.0045	.0030	.0010
1/4	.0065	.0040	.0025
1/2	.0080	.0050	.0030
3/4	.0080	.0050	.0030
1	.0090	.0070	.0040

Machinery's Handbook, Revised 21st Edition, Industrial Press Inc.

TABLE A-2   Machine and Cap Screw Sizes

Fastener		Nominal size	Decimal diameter	A	Clearance hole LMC*
Hex head machine screws		#8	.1640	.244	.204
		#10	**.1900**	**.305**	**.247**
		1/4	**.2500**	**.425**	**.337**
		5/16	.3125	.484	.398
		3/8	.3750	.544	.460
		7/16	.4375	.603	.520
		1/2	**.5000**	**.725**	**.612**
		5/8	.6250	.906	.765
		3/4	.7500	1.088	.919
Socket head cap screws		#4	.1120	.176	.144
		#6	.1380	.218	.178
		#8	.1640	.262	.213
		#10	**.1900**	**.303**	**.246**
		1/4	**.2500**	**.365**	**.307**
		5/16	.3125	.457	.384
		3/8	.3750	.550	.462
		7/16	.4375	.642	.540
		1/2	**.5000**	**.735**	**.617**
		5/8	6250	.921	.773
		3/4	.7500	1.107	.928
Round head machine screws		#4	.1120	.193	.152
		#6	.1380	.240	.189
		#8	.1640	.287	.226
		#10	**.1900**	**.334**	**.262**
		1/4	**.2500**	**.443**	**.346**
		5/16	.3125	.557	.434
		3/8	.3750	.670	.522
		7/16	.4375	.707	.572
		1/2	**.5000**	**.766**	**.633**
		5/8	.6250	.944	.784
		3/4	.7500	1.185	.967

*Clearance Hole Ø @ LMC is calculated with the formula (Fastener + Fastener head) / 2.

TABLE A-3   Number, Letter, and Fractional Drill Sizes

Drill no.	Fract.	Deci.	Drill no.	Fract.	Deci.	Drill no.	Fract.	Deci.	Fract.	Deci.
60	–	.0400	29	–	.1360	B	–	.238	**7/16**	**.438**
59	–	.0410		9/64	.1400	C	–	.242	29/64	.453
58	–	.0420	28	–	.1410	D	–	.246	15/32	.469
57	–	.0430	27	–	.1440		1/4	**.250**	31/64	.484
56	–	.0465	26	–	.1470	E	–	.250	**1/2**	**.500**
	3/64	.0469	25	–	.1500	F	–	.257	33/64	.516
55	–	.0520	24	–	.1520	G	–	.261	17/32	.531
54	–	.0550	23	–	.1540		17/64	.266	35/64	.547
53	–	.0595		5/32	.1560	H	–	.266	**9/16**	**.562**
	1/16	**.0625**	22	–	.1570	I	–	.272	37/64	.578
52	–	.0635	21	–	.1590	J	–	.277	19/32	.594
51	–	.0670	20	–	.1610		9/32	.281	39/64	.609
50	–	.0700	19	–	.1660	K	–	.281	**5/8**	**.625**
49	–	.0730	18	–	.1700	L	–	.290	41/64	.641
48	–	.0760		11/64	.1720	M	–	.295	21/32	.656
	5/64	.0781	17	–	.1730		19/64	.297	43/64	.672
47	–	.0785	16	–	.1770	N	–	.302	**11/16**	**.688**
46	–	.0810	15	–	.1800		**5/16**	**.313**	45/64	.703
45	–	.0820	14	–	.1820	O	–	.316	23/32	.719
44	–	.0860	13	–	.1850	P	–	.323	47/64	.734
43	–	.0890		3/16	**.1880**		21/64	.328	**3/4**	**.750**
42	–	.0935	12	–	.1890	Q	–	.332	49/64	.766
	3/32	.0938	11	–	.1910	R	–	.339	25/32	.781
41	–	.0960	10	–	.1940		11/32	.344	51/64	.797
40	–	.0980	9	–	.1960	S	–	.348	**13/16**	**.813**
39	–	.0995	8	–	.1990	T	–	.358	53/64	.828
38	–	.1015	7	–	.2010		23/64	.359	27/32	.844
37	–	.1040		13/64	.2030	U	–	.368	55/64	.859
36	–	.1065	6	–	.2040		**3/8**	**.375**	**7/8**	**.875**
	7/64	.1094	5	–	.2060	V	–	.377	57/64	.891
35	–	.1100	4	–	.2090	W	–	.386	29/32	.906
34	–	.1110	3	–	.2130		25/64	.391	59/64	.922
33	–	.1130		7/32	.2190	X	–	.397	**15/16**	**.938**
32	–	.1160	2	–	.2210	Y	–	.404	61/64	.953
31	–	.1200	1	–	.2280		13/32	.406	31/32	.969
	1/8	**.1250**	A	–	.2340	Z	–	.413	63/64	.984
30	–	.1290		15/64	.2340		27/64	.422	**1**	**1.000**

TABLE A-4 Inch to Millimeter Conversion Chart

Inches (Fractions)	Inches (Decimals)	Milli-meters	Inches (Fractions)	Inches (Decimals)	Milli-meters
	.00394	.1	15/32	.46875	11.9063
	.00787	.2		.47244	12.00
	.01181	.3	31/64	.484375	12.3031
1/64	.015625	.3969	1/2	.5000	12.70
	.0157	.4		51181	13.00
	.01969	.5	33/64	.515625	13.0969
	.02362	.6	17/32	.53125	13.4938
	.02756	.7	35/64	.546875	13.8907
1/32	.03125	.7938		.55118	14.00
	.0315	.8	9/16	.5625	14.2875
	.03543	.9	37/64	.578125	14.6844
	.03937	1.00		.59055	15.00
3/64	.046875	1.1906	19/32	.59375	15.0813
1/16	.0625	1.5875	39/64	.609375	15.4782
5/64	.078125	1.9884	5/8	.625	15.875
	.07874	2.00		.62992	16.00
3/32	.09375	2.3813	41/64	.640625	16.2719
7/64	.109375	2.7781	21/32	.65625	16.6688
	.11811	3.00		.66929	17.00
1/8	.125	3.175	43/64	.671875	17.0657
9/64	.140625	3.5719	11/16	.6875	17.4625
5/32	.15625	3.9688	45/64	.703125	17.8594
	.15748	4.00		.70866	18.00
11/64	.171875	4.3656	23/32	.71875	18.2563
3/16	.1875	4.7625	47/64	.734375	18.6532
	.19685	5.00		.74803	19.00
13/64	.203125	5.1594	3/4	.7500	19.05
7/32	.21875	5.5563	49/64	.765625	19.4469
15/64	.234375	5.9531	25/32	.78125	19.8438
	.23633	6.00		.7874	20.00
1/4	.2500	6.35	51/64	.796875	20.2407
17/64	.265625	6.7469	13/16	.8125	20.6375
	.27559	7.00		.82677	21.00
9/32	.28125	7.1438	53/64	.828125	21.0344
19/64	.296875	7.5406	27/32	.84375	21.4313
5/16	.3125	7.9375	55/64	.859375	21.8282
	.31496	8.00		.86614	22.00
21/64	.328125	8.3344	7/8	.875	22.225
11/32	.34375	8.7313	57/64	.890625	22.6219
	.35433	9.00		.90551	23.00
23/64	.359375	9.1281	29/32	.90625	23.0188
3/8	.375	9.525	59/64	.921875	23.4157
25/64	.390625	9.9219	15/16	.9375	23.8125
	.3937	10.00		.94488	24.00
13/32	.40625	10.3188	61/64	.953125	24.2094
27/64	.421875	10.7156	31/32	.96875	24.063
	.43307	11.00		.98425	25.00
7/16	.4375	11.1125	63/64	.984375	25.0032
29/64	.453125	11.5094	1	1.0000	25.4001

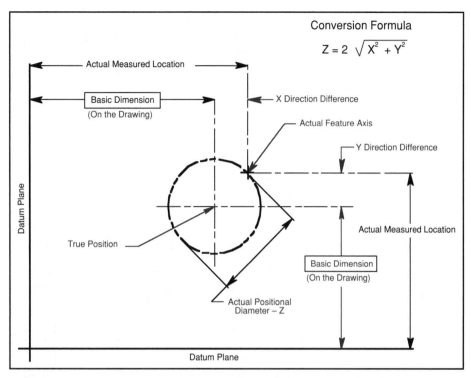

Figure A.1 Conversion diagram from coordinate measurements to cylindrical tolerance. (The Coordinate Conversion Table is on the next page.)

TABLE A-5   Conversion of Coordinate Measurement to Cylindrical Tolerance

Z Diameter Positional Tolerance

Y	.0301	.0303	.0306	.0310	.0316	.0323	.0331	.0340	.0350	.0360	.0372	.0384	.0397	.0410	.0424
.015															
.014	.0281	.0283	.0286	.0291	.0297	.0305	.0313	.0322	.0333	.0344	.0356	.0369	.0382	.0396	.0410
.013	.0261	.0263	.0267	.0272	.0278	.0286	.0295	.0305	.0316	.0328	.0340	.0354	.0368	.0382	.0397
.012	.0241	.0243	.0247	.0253	.0260	.0268	.0278	.0288	.0300	.0312	.0325	.0339	.0354	.0369	.0384
.011	.0221	.0224	.0228	.0234	.0242	.0250	.0261	.0272	.0284	.0297	.0311	.0325	.0340	.0356	.0372
.010	.0201	.0204	.0209	.0215	.0224	.0233	.0244	.0256	.0269	.0283	.0297	.0312	.0328	.0344	.0360
.009	.0181	.0184	.0190	.0197	.0206	.0216	.0228	.0241	.0254	.0269	.0284	.0300	.0316	.0333	.0350
.008	.0161	.0165	.0171	.0179	.0189	.0200	.0213	.0226	.0241	.0256	.0272	.0288	.0305	.0322	.0340
.007	.0141	.0146	.0152	.0161	.0172	.0184	.0198	.0213	.0228	.0244	.0261	.0278	.0295	.0313	.0331
.006	.0122	.0126	.0134	.0144	.0156	.0170	.0184	.0200	.0216	.0233	.0250	.0268	.0286	.0305	.0323
.005	.0102	.0108	.0117	.0128	.0141	.0156	.0172	.0189	.0206	.0224	.0242	.0260	.0278	.0297	.0316
.004	.0082	.0089	.0100	.0113	.0128	.0144	.0161	.0179	.0197	.0215	.0234	.0253	.0272	.0291	.0310
.003	.0063	.0072	.0085	.0100	.0117	.0134	.0152	.0171	.0190	.0209	.0228	.0247	.0267	.0286	.0306
.002	.0045	.0056	.0072	.0089	.0108	.0126	.0146	.0165	.0184	.0204	.0224	.0243	.0263	.0283	.0303
.001	.0028	.0045	.0063	.0082	.0102	.0122	.0141	.0161	.0181	.0201	.0221	.0241	.0261	.0281	.0301
	.001	**.002**	**.003**	**.004**	**.005**	**.006**	**.007**	**.008**	**.009**	**.010**	**.011**	**.012**	**.013**	**.014**	**.015**

Left axis: Y Direction Deviation

Bottom axis: X Direction Deviation

Index

ABOUT THE AUTHOR

Gene R. Cogorno has more than 25 years of experience as an educator and technical trainer. He spent 12 years developing and instructing a practical training program in industrial education, and taught machine technology at San Jose State University for 3 years. Mr. Cogorno joined FMC Corporation as a Senior Technical Trainer in 1984 and founded Technical Training Consultants in 1992. He earned his bachelor's and master's degrees in Industrial Education from San Jose State University.